Implementing Scalable CAN Security with CANcrypt

Authentication and encryption for CANopen, J1939 and other Controller Area Network or CAN FD protocols

by Olaf Pfeiffer

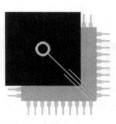

A technology guide from

EMBEDDED
SYSTEMS
ACADEMY

Jointly published by

Embedded Systems Academy, Inc.   Embedded Systems Academy GmbH
1250 Oakmead Parkway, Suite 210   Bahnhofstraße 17
Sunnyvale, CA 94085, USA   30890 Barsinghausen, Germany

Limitation of Liability

Available Editions

On its first release in March 2017, this book was made available in two editions:

Paperback, limited software version
ISBN 978-0-9987454-0-4

Black and white paperback edition. The demo software matching the examples in the book is available for download at www.esacademy.com/cancrypt. The software license of this software is limited to educational and evaluation purposes.

Hardcover, commercial software version
ISBN 978-0-9987454-1-1

Full color hardcover edition including a different, commercial CANcrypt software version delivered using a download code. The software license covers prototyping and an initial pilot production run of up to 500 devices.

Edition

This is the black and white softcover edition. The matching software demos can be downloaded from www.esacademy.com/cancrypt. This software is not suitable for commercial use, it only illustrates the main CANcrypt concepts.

About this book

When the Controller Area Network (CAN) was designed, security was not a requirement. The primary usage of CAN was considered closed; possible intruders or attackers would simply not get physical or remote access to the network. However, today it is more and more common that devices connected to a CAN system also have connections to other networks, including the Internet. Recent car hacks have shown that attackers may get access to CAN systems. Without strong security features, an attacker automatically gains full access to everything connected, allowing active control commands to be recorded and replayed.

In this book we examine which options developers of CAN based systems realistically can use to provide adequate security features.

What can we do...

- without introducing heavy-weight security protocols?
- to detect possibly injected messages?
- without any hardware change?
- with minimal software change and integration effort?

We introduce the open CANcrypt protocol and software interface, which provides a scalable and customizable CAN security system. Depending on the application requirements and resources available in the individual devices, various protection levels can be realized.

Acknowledgements

Writing a technical book is seldom a single-person achievement. I hereby express my gratitude to everyone who contributed to this book. Some did this knowingly, others unknowingly.

With his publications and talks, Klaus Schmeh (Schmeh) continuously keeps fueling my interest in everything related to cryptography and has done so for many years. His works and examples are a reminder that cryptography comes in many varieties.

The works of Ian Foster (Fast and Vulnerable: A Story of Telematic Failures, 2015) (Exploring Controller Area Networks, 2015), Charlie Miller (Miller, 2015), Chris Valasek (Miller, 2014), Craig Smith (Smith, 2016) and others have revealed various security vulnerabilities in systems using CAN. Their efforts have drastically increased the overall awareness for CAN-related security issues.

The NSA published the SPECK (Ray Baulieu, 2013) lightweight cipher in 2013. At first I was reluctant to use a NSA published cipher. However, until 2017 none of the many publications about it found any major caveat. A variation of SPECK is used for the default implementation of CANcrypt. However, just to be sure, its use in CANcrypt is on top of dynamically changing keys. An attacker who is able to determine a single dynamic key only has a few seconds (or milliseconds) to exploit it until the next key is used...

The CiA (CAN in Automation user's group) maintains most CAN and CANopen related standards. Participating in many standardization meetings has taught me that occasionally one just has to "do it," otherwise it might take years until first results are available.

My partners and colleagues Christian Keydel, Andrew Ayre and Ralf Kindermann offered continuous support and help with many of the technical aspects of the CANcrypt protocol as well as implementation-specific issues.

The lector for this book project has been Jan Axelson (www.janaxelson.com). Jan is a known technology writer and was a perfect match for this project from the start.

And last, book projects commonly require unusual working hours. And those require support and patience from the loved ones surrounding you. Leah, Magnus, Maria, thank you!

Olaf Pfeiffer
February 2017

Bibliography

Achieving Confidentiality Security Service for CAN. **Chavez, Rosete, Henriquez. 2005.** s.l. : Electronics, Communications and Computers, CONIELECOMP 2005, 2005.

CANAuth - A Simple, Backward Compatible Broadcast Authentication Protocol for CAN. **Anthony Van Herrewege, Dave Singelee, Ingrid Verbauwhede. 2011.** s.l. : ECRYPT Workshop on Lightweight Cryptography, 2011.

Checksum and CRC Data Integrity Techniques for Aviation. **Koopman, Philip. 2012.** s.l. : Electrical & Computer Engineering, 2012.

CiA 301. V4.2 2007. *CANopen Application layer and communication profile.* Nürnberg : CAN in Automation (CiA) e.V., V4.2 2007.

CiA 305. V2.2.17 2013. *CANopen Layer setting services (LSS) and protocols.* Nürnberg : CAN in Automation (CiA) e.V., V2.2.17 2013.

CiA 416. V2.0 2007. *Application Profile for Building door control.* Nürnberg : CAN in Automation (CiA) e.V., V2.0 2007.

CiA 447-1. V2.0 2012. *CANopen Application profile for special-purpose car add-on devices.* Nürnberg : CAN in Automation (CiA) e.V., V2.0 2012. Vol. 1: General Definitions.

Exploring Controller Area Networks. **Ian Foster, Karl Koscher. 2015.** 6, s.l. : login www.usenix.org, 2015, Vol. 40.

Fast and Vulnerable: A Story of Telematic Failures. **I. Foster, A. Prudhomme, K. Koscher, S. Savage. 2015.** Washington, DC : Proceedings of the 9th USENIX Workshop on Offensive Technologies , 2015.

Koopman, Philip. Best CRC Polynomials. [Online] Carnegie Mellon Univeristy. https://users.ece.cmu.edu/~koopman/crc/.

Miller. 2015. *Remote Exploitation of an Unaltered Passenger Vehicle.* Las Vegas, NV : Black Hat USA, 2015.

Miller, Valasek. 2014. *A Survey of Remote Automotive Attack Surfaces.* Las Vegas, NY : Black Hat USA, 2014.

Pfeiffer, Ayre, Keydel. 2003. *Embedded Networking with CAN and CANopen.* San Jose : s.n., 2003.

Plug-and-secure communication for CAN. **Andreas Mueller, Timo Lothspeich, Robert Bosch GmbH. 2015.** Vienna : international CAN Conference, 2015.

Ray Baulieu, Douglas Shors, Jason Smith. 2013. *The SIMON and SPECK Families of Lightweight Block Ciphers.* s.l. : NSA, 2013. Cryptology ePrint Archive 2013/404.

Schmeh, Klaus. Klausis Crypto Kolumne. [Online] http://scienceblogs.de/klausis-krypto-kolumne/.

Smith, Craig. 2016. *The Car Hacker's Handbook: A Guide for the Penetration Tester.* s.l. : No Starch Press, 2016.

Voss, Wilfred. 2008. *A Comprehensible Guide to J1939.* s.l. : Copperhill Media, 2008.

Contents

This is an overview of the main chapters and headings.
For a full table of contents, see Appendix E at the end of the book.

"Encryption works. Properly implemented strong crypto systems are one of the few things that you can rely on"

"Dawned Downers"

1 Introduction

The CAN (Controller Area Network) is a 30-year-old communication technology that worked well for much of its history without any security features. So what changed that now some 30 years later we need to review the network's security features?

The original use case for CAN systems was that of a closed network. The network would be deeply embedded in machinery without any connection to other networks or the Internet. In this case, any hacker attack would only be a physical attack – to get access to the CAN subsystem, a hacker would need physical access to the machine.

However, today the CAN subsystem is no longer self-contained. More and more often, bridges and gateways to other networking technologies are added, including connections to devices that have access to the Internet. Some examples of these devices are remote access devices for diagnostics or maintenance and multimedia servers like those in use in some automotive applications.

When we add devices that implement a gateway to the Internet or offer wireless communication options like Bluetooth and Wi-Fi, we also open a door for possible attacks to the CAN network.

Once a possible intruder is past the firewalls that limit access, there are typically no further hurdles as all communication is unprotected and in the case of CANopen or J1939, even well documented.

1.2   Prerequisites

You should have a reasonably good understanding of how the Controller Area Network works before continuing reading. If you are unsure, we suggest reading (CiA 301, V4.2 2007) or one of the many online "Controller Area Network Primer" documents like
www.computer-solutions.co.uk/download/Peak/CAN-Tutorial.pdf

Embedded Systems programming knowledge is required as soon as you want to start implementing CANcrypt on microcontrollers. In this book you will find pseudo code as well as a description of the C functions that make up the user interface of CANcrypt.

1.3   What is at risk?

Some might ask what the specific security risk is. Here the popularity of CAN comes into play. These days almost everything with wheels uses CAN (including electric bikes). Maritime and avionic use is also common. You can find CAN in elevators, medical equipment (including surgery robots), and many industrial control processes. Various "car hacks" have been made public, including odometer manipulation and unlocking doors but also active control of steering and brakes.

CAN is also a popular communication channel for software updates. Often new software can be loaded into components of a system via a CAN channel. This ability is a hacker's dream come true – being able to load software into a device with an embedded system that otherwise would not be accessible.

If these systems can be hacked, and hackers can both read and actively send commands/control, then we have to ask ourselves:

- How many hackable vehicles, ships and planes are out there?
- How many hackable elevators are out there?
- How many hackable medical devices are there?
- Is it possible to manipulate any of the above in a way that someone gets harmed and it is only recognized as "technical malfunction"?

Today some of these questions might still sound like the topic for a Hollywood blockbuster. However, it makes one wonder how Edward Snowden would evaluate the likeliness of such scenarios. Maybe it has already been done and just not published...

1.3.1 Known hacks

Various papers and articles about car hacks have been published, and the hacks often involve CAN communication. A good summary of recent activities is in the article (Exploring Controller Area Networks, 2015) and in (Fast and Vulnerable: A Story of Telematic Failures, 2015).

Where some car hacks focus on manipulating the odometer, others have successfully taken remote, active control of park-assist functions to actively steer and brake (Miller, 2014) (Miller, 2015). Other hacks were able to read out software or firmware parts and alter them.

In summary, such vulnerabilities are scary. From an engineering viewpoint it is sad that most of these hacks are possible because there is no security functionality implemented at all. So far, the security has been focusing too much on the firewall aspect – protecting access to the CAN network in the first place. However, after overcoming this barrier, an intruder often faces no additional barriers.

1.4 Related work

CAN security is not a new topic. It has been addressed in several conferences, reports, and papers for more than 10 years (Achieving Confidentiality Security Service for CAN, 2005). In 2007, the CiA (the CAN users and manufacturers group) standardized encryption in an application profile (CiA 416, V2.0 2007). Yet solutions are not commonly present in today's applications. Why is that?

There seems to be a gap between what is academically possible and what is practical and easily available.

For example, any solution that requires a hardware change of the currently used CAN controllers is simply not practical for broad market applications. Chip manufacturers do not embrace new CAN communication technologies easily – we can see that it takes years until something new like CAN FD is widely adopted. Examples of such hardware-oriented solutions include (CANAuth - A Simple, Backward Compatible Broadcast Authentication Protocol for CAN, 2011) and (Plug-and-secure communication for CAN, 2015).

Engineers with an interest in adding security to a CAN system can easily find multiple papers addressing the topic. Often highly specialized, these papers may focus on automotive applications, authentication without encryption, broadcasts, or node pairing. The CANopen application profile for building doors (CiA 416, V2.0 2007) features security options, but the majority of CANopen users are not aware of this application-specific document.

Without being an expert in security, which system or method should we choose? How much security do we really need in broad market applications? And where do we start when it comes to actually programming?

Just to name one example: a function required for authentication is either a secure hash algorithm (SHA) or a checksum or CRC that is encrypted. As commonly used SHAs have block and output sizes far bigger than our data, we could assume a simple checksum selection can do the trick. However, selecting an appropriate checksum or CRC method is a science of its own. There are methods like Adler and Fletcher checksums as well as the cyclic redundancy checksums where the art is in selecting an appropriate polynomial (Checksum and CRC Data Integrity Techniques for Aviation, 2012) and (Koopman).

Our CANcrypt approach not only provides a generic, low-weight CAN security system that is protocol independent at the application and higher layers. We also ensure that there is room for customization. Depending on the application's requirement, we support a variety of security methods, from minimal protection using minimal resources to customizable methods that allow higher-end encryption and authentication methods. Full C source code demo examples are provided, simplifying adaptation and integration.

The system supports protocols to generate, store, or erase a hierarchy of keys – these functions can be done by the manufacturer or by a system integrator or system owner, each having access at their own security level. Add-on solutions for key management are provided.

1.5   Key management

Another reason why manufacturers are probably reluctant to broadly implement security features is the issue of proper key management. This is also one of the biggest challenges of the "Internet of Things" discussion.

No matter which key type is used, synchronous (same key shared) or asynchronous (public and private key), some sensitive (synchronous or private) keys need to be installed onto the embedded devices at some point. Who is doing it and when? Where and how are these keys generated? Will there be copies of the keys on a PC? Imagine there is one PC on a production line that is in charge of installing millions of keys for embedded devices. What an interesting target for hackers...

CANcrypt provides functionalities that simplify key management. The main feature introduced in this respect is the ability for a pair of nodes to randomly generate or exchange a key between themselves – invisible to the other network participants. If the key is stored locally, no copy of the key needs to exist anywhere else.

A second feature is that CANcrypt supports multiple keys in a hierarchy, for example, one each installed by a manufacturer, a system integrator or machine builder, and the system owner. Keys lost or erased (as is typically the case when single devices are repaired/replaced) can be re-generated on the next higher hierarchy level.

And last, CANcrypt allows keys to be combined with other device-specific elements such as a local serial number. Thus a manufacturer does not necessarily need to install different keys on each device. Devices can use a shared key if other distinguishing elements like a product code, revision number and serial number are available.

1.6   What about safety?

This book's primary goal is to address the security aspects of authentication of data as well as data encryption. Although aspects for safety-relevant or safety-critical communication are not a focus, the basic methods introduced in this book are well suited to be used to cover safety issues as well. For both  safety and security implementations, we need to be able to detect and react to failures. The main difference between safety and security is that when we talk about safety, we do not necessarily assume that failures are intentional. On the other hand, when talking about security, we assume that failures might be provoked by attackers and intruders.

Whether bits and messages are manipulated on purpose or randomly, we need to deal with it. And to a certain extent the methods used can be applied to both security as well as safety aspects. In the appendix we summarize which CANcrypt security aspects introduced in this book can also be applied for safety applications.

1.7   Usage with I²C, RS-485 and others

Many of the methods introduced in this book can also be applied to other lightweight embedded networking technologies. However, one of the central elements of CANcrypt is the bit-generation cycle. Here two nodes can generate or exchange a bit without other nodes being able to detect which bit was generated or exchanged. For this method to work, a shared transmission medium is required (not dedicated transmit and receive lines). The shared medium could be a single shared wire or pair of wires, as in RS-485, or the I²C data line in multi-master mode.

If we see enough requests, we will adapt CANcrypt to other suitable communication technologies in the future.

1.8   Definitions and Terminology

Cancrypt uses data types as defined in CANopen. Here are some of the common terms used:

address, device
Each device has a unique address. This is comparable to the node ID in some network technologies. CANcrypt supports up to 14 devices using the addresses 1 to 14.

bit-generation cycle (or bit claiming)
Two nodes can exchange a bit without others on the network being able to detect the bit value. This process, which requires multiple messages, is the bit-generation cycle.

CAN message ID

Each message has a unique identifier called the CAN message ID, typically 11 bits, sometimes 29 bits (referred to as an extended identifier).

configurator, CANcrypt configurator

The configurator actively configures individual devices for functions such as key generation, assignment, storage, and erase. The configurator is not required during regular operation and has the CANcrypt address 15.

device, CANcrypt device

A system may have up to 14 devices capable of processing CANcrypt messages.

error counter

Each device or configurator maintains an internal error counter. The count is incremented with every suspicious system behavior, including errors. If the error counter reaches a specified value, secure communication is halted.

grouped, grouping

Every active device is in a group where all devices share a common dynamic key and communicate with each other securely. Messages are authenticated and optionally encrypted and decrypted based on the group key.

index, to data object

All parameters are addressable using an index and sub-index value. This addressing method is adapted from CANopen and other networking technologies.

INTEGERxx, data type

INTEGERxx is the notation for a signed integer value where "xx" indicates the number of bits used by the data type. CANcrypt uses "little endian" notation.

key, dynamic

A dynamic key is the base for all cryptographic functions. This key gets updated frequently and synchronously by all active devices.

key, hierarchy

Implementing of a key hierarchy is recommended to enable a device to have multiple keys with different levels of authority. For example, a manufacturer key might give access to a bootloader, while a system builder key might give access to system configurations.

key, management

As soon as keys get permanently stored in devices, key management is required to determine when each key is generated and by whom. CANcrypt supports multiple models of key management.

key, permanent

At least one key must be permanently stored in each device. If a device supports storing multiple keys in a hierarchy, the term permanent key refers to the stored key that is currently in use.

message, secure

CANcrypt transmits each secure message paired with a preamble message that announces the secure message that follows. The secure message contains all of the data in the original (unsecure) message, possibly encrypted.

message table

The message table is shared by all active devices. It lists all CAN messages that require security handling. A device that receives a CAN message listed in this table may pass the message to the application only if the message was received with the matching preamble and signature. For details see section 5.4 CANcrypt secure message table.

one-time pad, pseudo

For all encryptions and authentications a pseudo one-time pad is used. This one-time pad is available to all grouped or paired devices so that it can be used like a synchronous key.

paired, pairing

Two active devices may get paired to have an individual secure communication channel. Messages are authenticated and optionally encrypted and decrypted based on the pairing key.

preamble

The preamble is a CAN message that announces the secure message that follows. The preamble contains control data as well as a signature that covers the control data and the secure message.

STRING4, data type

STRING4 is the notation for a string segment containing four ASCII characters. The string is zero-terminated. All unused bytes of this string are set to zero.

sub-index, to data object

All CANcrypt parameters are addressable via index and sub-index values. This method to address data is adapted from CANopen and other networking technologies.

UNSIGNEDxx, data type

UNSIGNEDxx is the notation for an unsigned integer value where "xx" indicates the number of bits used by the data type. CANcrypt uses "little endian" notation.

"The secret of getting ahead is getting started."

"Martin Wak"

2 Security for CAN systems

As mentioned in section 1.2 , one of the prerequisites for reading this book is some CAN knowledge. The core features of CAN, at-a-glance, are:

- Messages have a unique message identifier. There are several CAN protocol versions:
  - CAN 2.0A uses 11-bit identifiers
  - CAN 2.0B (extended) uses 29-bit identifiers
  - CAN FD (flexible data rate/field)
- The message ID is used for bus arbitration; the lower value wins.
- The data size is:
  - CAN: up to 8 bytes
  - CAN FD: up to 64 bytes
- The data overhead per message is 6–11 bytes
  - 11 bits: start/end of frame and intermission
  - 11 or 29 bits: message identifier
  - 6 bits: various control bits
  - 4 bits: DLC (data length code)
  - 15 bits: CRC
  - 0–xx bits: bit stuffing (roughly 1bit per byte)
- Error detection is built in with automatic retry on CRC failure.
- Maximum bit rates are:
  - CAN: 1 Mbps
  - CAN FD: uses a higher bit rate during the data field, can be a multiple of the regular bit rate
- All messages are broadcast, every connected device receives everything.

2.1   CAN security challenges

Security in embedded systems must never focus on just one subsystem and should always offer multiple levels of protection. Adding security to CAN helps reduce risk. However, it's best to prevent hackers from getting access to the network in the first place. So any gateway to other networks should always have a firewall that prohibits unauthorized access to the neighboring network.

As mentioned above, a hacker that has access to a CAN system can fully stop all operation (DOS attack). However, that type of attack typically results in a safe shutdown of the system, immediately recognized by any user or operator.

What we need to achieve is secure data and commands that cannot be read or manipulated. While a CAN system is operating, we want to be reasonably sure that data is authenticated, encrypted, and not manipulated. Doing so typically requires inserting extra messages that contain the security information required, such as a signature.

2.1.1   Data sizes and content

One challenge with adding security to CAN systems is that the data communicated is often very small. A command to switch something on or off can be a single bit.

Also, the data might not change often. The same command byte may transmit over and over every 100 ms and not change for minutes at a time. This type of data is especially difficult to encrypt if you also need to ensure that the secure version must change with every transmission. (Otherwise a simple record and replay of messages could be used successfully as a hacker attack.)

Therefore a CAN security system should support encryption only of individual bytes. Data that remains the same most of the time should not be encrypted as this would provide an intruder with an attack vector if parts of the CAN data are known. For example, in CANopen some service transfers always start with the same command byte – here, encryption should only be applied to the data field.

2.1.2   Communication via broadcasts

At the lowest level, all CAN communications are broadcast. Multicasts or point to point communications are realized by filtering (effectively ignoring) messages on the receiver side. Every device connected to the network receives every data message. Typically the message ID (or some of the data contents) introduces a source and destination communication model. A specific message ID could be reserved to transmit from one specific sensor to one specific controller with all other devices on the network ignoring the message.

Some messages are multicast from one node to a selected group. For example, a temperature sensor might send data to multiple devices.

Multicast and broadcast present an additional challenge to security. Keys and methods must be synchronized not just between a pair of devices but by all devices that need to be able to receive the secure multicast or broadcast.

2.1.3 Microcontroller Resources

Many microcontrollers in CAN devices have limited memory and processing power. Security mechanisms that are common on the Internet often require thousands of operating cycles for authentication, encryption, and decryption and require that both the key and data are a minimum size, often 256 or more bits. These requirements aren't practical for microcontrollers with limited resources.

2.2 Attack vectors

A CANopen-connected device can be vulnerable to a number of possible attack scenarios. A device that is also available on the Internet could be hijacked from there. A similar attack scenario would be if a CANopen-connected device is also available via Bluetooth or WiFi and unauthorized access is made from there. Probably the most unlikely attack is physically hooking up a sniffing device directly to the CANopen network. However, as we are thinking about how to make a network more secure, we can take this attack option into consideration, too.

On the bottom line, all of these attacks mean that intruders can see all CANopen messages communicated and can also inject messages. As all communications are unprotected and mostly unmonitored for harmful injections, systems are easy to manipulate after obtaining access at this level.

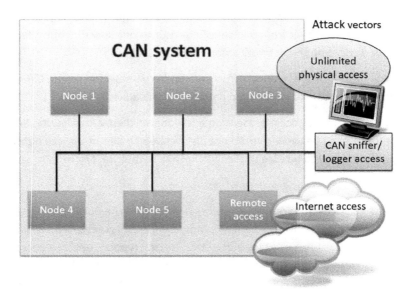

CAN ATTACK VECTORS

The protection options vary with the CANopen device hijacked by the intruder. If the device is the main master or manager of the system, detection might not be possible as there might be no "unusual" CANopen traffic. However, if the hijacked device is a regular CANopen device, then depending on what the intruder does, there can be ways to detect that additional, unexpected messages have been injected.

The figure above ("CAN Attack Vectors") illustrates the three typical security threat levels a CAN system might get exposed to: unlimited physical access, sniffer access, and remote (Internet) access.

2.2.1 Unlimited physical access

From the physical point of view, many CAN systems are still closed. They are embedded within machinery, and access to the wires or individual devices is limited. If an intruder can unplug and replace devices or cut the wires and place a bridge or gateway into the system (trying a man in the middle attack), simpler security levels might be easy to overcome.

If an intruder has unlimited physical access to the entire network including device PCBs, the available security options are very limited. Having access to all debug ports of the microcontrollers provides many other attack vectors besides manipulating CAN.

Often physical access is only possible through a violation of trust, when someone who was trusted with access to the system willfully violates that trust. Not all CAN security implementations need to cover this scenario.

2.2.2 Sniffer access

An intruder that has direct access to a CAN system (by connecting a sniffer device or a laptop with a CAN interface) can read ALL communication on the network. With write access, "denial of service" style attacks (swamping the bus with messages so nothing else gets through) are easy and cannot be prevented by any security mechanism added.

2.2.3 Remote access

The last attack vector is remote access through a device that is a gateway to other networks. Recent hacks have shown that this sort of access is not just theoretical. With enough time and effort, hackers can get access to such systems, monitor all communications, and even inject control commands.

Note that today manufacturers can no longer guarantee that a CAN system they deliver will always remain closed. CAN devices with wireless gateways like Bluetooth or even WiFi are readily available on the market, and CAN remote access solutions are offered by multiple suppliers.

2.3 Selected security use cases

As we will see in the next chapter, security for CAN comes at a cost. Resources like memory, processing power, and extra bandwidth are required. Especially in CAN systems, these resources can be very limited so it most likely will not be feasible to add security to all data communicated. It is worthwhile to review some selected use cases in regard to which data requires security.

2.3.1 Activating manufacturer specific functionality

Many CAN systems implement enhanced manufacturer-specific functions that can be activated via CAN commands. These functions can include things like additional diagnostic channels or advanced configuration functionality. As these are often system critical, all such communication should be both authenticated and encrypted.

2.3.2 Bootloading / loading software blocks

Whenever a device can be reprogrammed or can load software, the entire process, from activating the mode that enables reprogramming as well as the actual transfer of the software, should be both authenticated and encrypted.

2.3.3 Anything moving

If we look at vehicles of any kind, including swimming or flying, anything resulting in an active change of the controls (accelerating, breaking, steering) should be authenticated. Such commands should only be accepted from authenticated communication partners. The commands do not need  to be encrypted unless you do not want an intruder to be able to see what is currently happening.

In regard to encryption, you would probably want to secure all data that a position can be derived from; otherwise an intruder could easily track the vehicle. This data includes satellite navigation data as well as odometer or wheel pulse counter values.

2.3.4 Building control

Any command actively controlling an elevator, temperature, or light should be authenticated. A CANopen standard exists for locking and unlocking doors (CiA 416, V2.0 2007). However, the standard did not specify management of the keys , and to simplify maintenance, a simple backdoor has been integrated into the standard, allowing erasure and resetting of keys.

Burglars would be interested to know if someone is in a residence or if it is empty. Data that can be used to make a prediction about someone being present or not should be encrypted.

2.3.5 Industrial manufacturing processes

Industrial control applications vary a lot, and what to protect and how to protect it depends on multiple factors. For example, imagine a CAN system that steers valves that control the components of a secret recipe. An intruder with full CAN access would now be able to learn the recipe – and might be able to manipulate the manufacturing process so the end product does not have the desired quality. Such things have already be done by targeted viruses like Stuxnet.

In industrial manufacturing, one needs to carefully evaluate on a case-by-case basis which data should be protected and how.

2.3.6 Medical equipment

CAN networks are used in hospital beds, dialysis systems, scanners, and even surgery robots. On one hand, security should be added whenever a patient's health is at risk. However, as it might not be feasible to add both authentication and encryption to all data, one should take extra precautions in regard to additional security levels. Devices should remain "closed" as much as possible with remote access and any Internet connection reduced to an absolute minimum with strong tests of the firewall(s). Physically installing a network sniffer should not be easy.

2.4 Protection levels: desired or required?

Looking at the applications, we can identify a broad range of security requirements. On the simplest security level, intruder detection as explained in chapter 3 "Minimal precautions to take" might be sufficient. Here a system is considered secure until unexpected messages arrive and trigger an alert. This basic mechanism should be included in all higher security levels as well.

However, this mechanism fails if the intruder manages to disconnect a device or switch it offline or silent (for example, by setting the device into test or bootload mode).

An intruder might also cause a device to disconnect by provoking collisions. If a device transmits a message at regular intervals, an intruder can time collisions by transmitting messages with the same CAN ID at the expected transmit times of the device's messages. An attacked device that repeatedly detects a collision in the data field will eventually go "bus off" – disconnecting itself from the network.

Once a device is offline, an intruder could try to replace the device with its own. If the replacement device looks OK to the other devices, the devices will accept commands from it.

The next higher levels of security need an authentication mechanism to prevent intruders from injecting critical messages. One method of authentication uses two messages, one containing the data and the other containing a signature that authenticates the data. Another option is the secure heartbeat method described in section 2.4.1 .

To provide different levels of security, a CAN security system needs to support multiple kinds of checksums and signatures. Not all applications can support or need high-end hash functions, which can require many resources (processor cycles or memory). CANcrypt supports both 16-bit and 32-bit checksums as well as multiple (including customizable) algorithms.

Encryption should only be used selectively as outlined in the application examples above (Selected security use cases"). Again, a CAN security system needs to be flexible enough to support multiple methods because the highest end cryptographic functions will not be suitable for all applications.

2.4.1   Authentication with a Secure Heartbeat

If the main security function required is authentication and if multiple messages require authentication, a method requiring less overhead is a secure heartbeat. This is a cyclic message containing the current security status and a signature. With this method, a device does not authenticate every CAN message it transmits but instead authenticates only its heartbeat.

In such a setup, every device participating in authenticated communication must:

a)   monitor the network for any injected messages (detect messages that use a CAN ID that the device has used for transmit)
b)   while no injection has been detected, transmit the secure heartbeat
c)   when an injection is detected, transmit an alert and stop the secure heartbeat

On the receiving devices, we see:

a)   a secure heartbeat
if all previous data received from the device is authenticated
b)   an alert
if previous data was manipulated
c)   a secure heartbeat timeout
if previous data was manipulated or the device was taken offline

Systems using this secure heartbeat must carefully select the heartbeat repetition time. All data transmitted since a previous heartbeat is authenticated only by receiving the next heartbeat. If the secure heartbeat timeout is 1 second, detection of a timeout means that all data received within the last second may have been manipulated.

A secure heartbeat can run at a higher frequency, such as 100 ms, but a higher frequency uses more bandwidth and would need a shorter timeout period. We suggest a factor of 1.5, or 150 ms for a 100-ms heartbeat.

In the end, the specific application determines the acceptable timeout. Even 1 second might be acceptable if the controlled process is slow, like a heating control.

2.5   Differences between CAN and CAN FD

In 2014, ISO CAN FD (Flexible Data), a modification that increases data size and throughput, was introduced.

2.5.1   CAN

The traditional CAN system in use for over 30 years is based on a message system that uses an 11-bit CAN ID (message identifier) and up to eight bytes of data.

There is a version using the so-called extended identifier where the CAN ID is 29 bits. This version is used by some higher layer CAN protocols like J1939.

Note that we do not want to modify or encrypt the CAN ID as it is essential for routing messages (source and possibly destination).

Data sizes in CAN range from one byte (no chance to protect messages without data) to eight bytes (64 bits). Note that protection might not be required for the entire contents of a message.

2.5.2   ISO CAN FD

In CAN FD (Flexible Data), the maximum data size of a single message has been increased to 64 bytes, so here we need to be able to protect data sizes from one byte to 64 bytes (512 bits).

Also with CAN FD, the protection required might be limited to single bytes within a message.

Due to the increased data size, some of the more traditional encryption methods and algorithms might be used. However, a system that has been established for CAN can most likely also be used for CAN FD.

Note that CAN FD controllers are incompatible with regular CAN controllers. This means CAN FD can only be used if all devices connected are CAN-FD capable.

2.6   Higher-layer protocols

When we talk about networking technologies, we also talk about layers. Network communication is organized in layers, with a physical layer at the bottom followed by a data link layer, multiple protocol-specific layers, and the application layer on top.

CAN is considered to partially implement the physical and data link layers. Many protocol stacks are available that run on top of CAN. Among the more popular such protocols are CANopen and J1939.

Any security-level build for CAN should be positioned above CAN but below the chosen protocol stack so the code does not need to be modified for every proto-col stack. This arrangement is a bit different from Internet networking technolo-gies, where the security layers typically sit at higher levels.

The CANcrypt technology presented in this book sits directly above the CAN layer and can be transparently used with multiple protocol layers on top of CAN.

However, if the security layer is used unwisely with the protocol layers above it, an attacker might find it easier to hack the system. For example, if there is a regu-lar heartbeat message that mostly contains the same known data (because it is specified by the standard used), it would not be wise to add security to that mes-sage. A hacker could start guessing the value and use that as a starting point for a decrypting attack.

2.6.1   CANopen

CANopen (CiA 301, V4.2 2007) is a popular protocol used for many different appli-cations including industrial controls, medical devices, building control and eleva-tors, special purpose vehicles like police cars and taxis (CiA 447-1, V2.0 2012) and e-bikes.

The challenging parts of security for CANopen include a powerful NMT (Network Management) master broadcast. One such message can simultaneously signal all connected devices to leave operational mode. The damage caused by injecting an unauthorized NMT message is roughly comparable to a DOS attack where every-thing stops.

CANcrypt provides two options to add security for the NMT master message. The NMT Master can produce a CANcrypt secure heartbeat during normal operation and produce a CANopen emergency message on detecting an injection. A continuing secure heartbeat confirms to the receivers that no injection has happened. However, due to delays in transmit and receive FIFO handling, 100% confirmation requires receipt of the second secure heartbeat after an NMT message.

The second option is to add the NMT master message to the list of secure messages. The NMT master would then add the CANcrypt secure preamble to each NMT master message. However, this requires that ALL devices connected run the CANcrypt secure messaging system.

CANopen also offers layer-setting services (LSS) (CiA 305, V2.2.17 2013), which allow changing the node IDs of connected devices. If enabled, these services could allow an intruder to give a node a different node ID and then try to fake being the replaced node.

The security challenge here is that LSS is often used when nodes do not yet have a node ID and likely do not yet have a CANcrypt address either. If the communication is used only for setup or configuration of nodes, a default CANcrypt address might be a solution. Otherwise, a case-by-case examination of how an application uses LSS can help decide the best security strategy.

Another transfer mode possibly requiring special treatment is the segmented or blocked SDO (Service Data Object) transfer. This mode uses multiple CAN messages to transfer larger data blocks. Here, adding security overhead to every message might not be suitable. A possible approach is again self-monitoring of a node's own messages in combination with the secure heartbeat. As these are longer transfers, waiting for the following heartbeat(s) as confirmation should be acceptable.

2.6.2   SAE J1939

The Society of Automotive Engineers maintains the J1939 specification, which is popular for larger diesel engines, off-road construction machinery, and agriculture machinery. The specification uses CAN V2.0B with 29-bit message identifiers. CANcrypt does not distinguish between 11-bit and 29-bit CAN message identifiers, so no special treatment is required here.

In regard to communication messages, J1939 does not use a global command like the CANopen NMT master message, so there is one less security challenge to handle.

A specialty of J1939 is its node address claiming procedure at startup. Right after startup, J1939 devices do not yet have a node address, and since this will normally mean they also do not yet have a CANcrypt address, security methods cannot be used and must wait until all devices have claimed their node addresses.

As long as all devices have non-volatile memory to store keys and configurations, a security solution is possible if all participating devices boot up at the same time, their roles do not change, and no devices are removed or added. For such cases, CANcrypt provides a pairing/grouping mechanism.

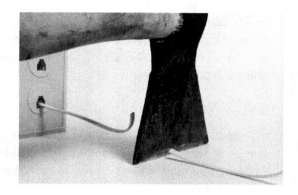

3 Minimal precautions to take

In this chapter we examine how to use existing CAN and CANopen mechanisms and features to detect when an intruder tries to actively manipulate a network. Even though CAN-based communication has no security built in, we can take precautions to minimize the impact of an intruder. Even if the intruder can passively read all data, we can at least detect if there are active manipulations going on in the system and possibly ignore fake, injected commands.

The following guidelines help making CAN and CANopen networks more secure and less vulnerable to attacks.

3.1   Avoid CAN message configurability

Flexible systems allow the reconfiguring of CAN messages used during operation. These changes can include things like triggering or timing of messages as well as which CAN ID is used for what purpose.

A CANopen example is the Process Data Object (PDO, mainly a CAN message with process data). In general, CANopen allows PDOs to be reconfigured via the CANopen network. By sending just a few messages to a device, PDOs can use different CAN message identifiers, different timing, and if supported, even different data contents.

If these parameters can be changed, an attacker with full network access can reconfigure which CAN messages a device listens to. A change could be made so important devices listen only to messages and commands coming from the attacker.

To increase security, such message configurations should be constant or read-only so they cannot be changed during operation. In CANopen, this security measure is advised not only for PDOs, but for all configurable CAN message types like SYNC, EMCY and SDO messages.

3.2 Avoid change-of-state (COS) triggering

In order to easily detect attacks, all communication should be very predictable so unexpected communication can be detected easily. If a message is triggered by COS only, injected messages cannot be easily detected as they can occur any time. However, if a message is transmitted on a fixed time slice or in response to another message (like a SYNC signal), injected messages can immediately be recognized as they are unexpected. A message that happens to be sent at exactly the same time as the original will cause an error frame.

3.3 Advanced monitoring of messages received

If a CAN message is time-triggered or SYNC-triggered, the device receiving the message should monitor if the message is received as expected or if it is an unexpected receive. Something that is unexpected (not in response to a SYNC or outside the expected time window) could be a first step toward detecting an injected message from an intruder.

In general, a single occurrence must not trigger the biggest alarm possible. Preferably, the event increments something like a "suspected injection counter". If the counter hits a certain number, the device should transmit some sort of an emergency message informing everyone that an intruder might be present. Alternatively and depending on the application, the device might also consider not accepting or processing the data.

3.4 Listen for own transmit CAN message identifier

All devices should listen for all CAN message identifiers that they use for transmitting data. In CAN and CANopen, all CAN message identifiers (with a few exceptions) are assigned to exactly one device – only that device may transmit a message with this ID. Therefore, a device that listens to the CAN IDs it transmits can immediately recognize if someone else tries to communicate using the same ID. A device detecting such an event should then generate an appropriate emergency message to inform everyone else on the network of the possible intruder (or wrong configuration of a device).

3.5   Physical separation of networks

A more drastic option to increase security is the physical separation of sensitive parts of the network using a smart bridge or gateway as a firewall. A bridge or gateway can separate parts of the network to limit the attack vectors available to intruders. With this strategy, access to one part of the network does not immediately include access to *all* parts.

Note that on a single CAN bus, all communication to and from every device is visible. A smart bridge or gateway configured as a firewall does not pass on every CAN message but instead implements filters to pass only those messages that the receiving side requires.

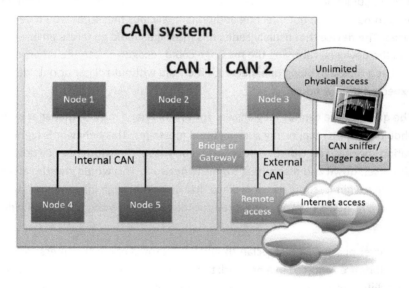

A BRIDGE OR GATEWAY AS FIREWALL

As an example, imagine that some machinery or vehicle using CAN has some parts of the network wiring hidden inside and some parts exposed externally. What are the options for physical access to the wiring to connect a sniffer or other remote access or manipulation device?

An intruder will find it much easier to connect something to the external wiring without getting noticed. Opening the machine or vehicle will draw much more attention and might require lock picking.

If the external part of the CAN system is not directly wired to the internal part but instead has a bridge or gateway in between, the functionality available on the external part can be drastically reduced. Filters in the bridge or gateway can be set to pass only those CAN messages that are expected and needed. The firewall can detect unknown or unexpected messages and initiate an alarm.

For an intruder, access to such a network no longer includes full access to everything and thus becomes less interesting for use as an attack vector.

3.6   Summary

If these guidelines are implemented, a potential attacker might still be able to see all communication, but injecting additional malicious messages would be recognized. The device that usually sends this message would generate an emergency, and the device(s) receiving the message could recognize both the emergency as well as a message being unexpected if it occurs without trigger or outside the expected time window.

The question we cannot fully answer for you is what a master/manager device should do upon recognizing such injected messages. This behavior is highly application specific. The worst reaction would be that an individual device decides to shut itself down and disconnect from the network. This would give the attacker a chance to mimic, simulate, or replace that device through injected messages. So a shutdown should only be considered for the entire network, not for individual devices.

Wherever possible, a technician should be informed of injected messages. If such injections are reported in a network that previously ran for weeks or months, then definitely something is seriously wrong. The network is either under attack or has some other major issues (like duplicate node ID) and needs to be checked anyway.

One side effect is that active monitoring or analysis tools (that also transmit messages) can no longer be used in such a system. Only passive monitoring can be used. If you think a solution is to put devices into a special test mode that allows active monitoring (and the security monitoring introduced above is disabled), then the hackers have won. Their first goal would be to enable these test modes before actively injecting further messages.

In summary, the guidelines above are easy to implement as typically all basic mechanisms described are already in place within the code of exiting CAN or CANopen implementations. On the downside, the additional protection is limited because injected messages cannot be directly identified. We can only recognize that a possible injection is taking place.

"If you want to keep a secret, you must also hide it from yourself."

"Roger Gleelow"

4 Selecting cryptographic methods

Many CAN devices are based on microcontrollers with limited memory and processing power. Yet at the highest supported speed of 1 Mbps, the CAN message rate can be as high as 10,000 messages per second. At this rate, adding reasonably safe security software to existing devices may be a challenge.

As in most security systems, there is a tradeoff between how much security we need vs. how much we can afford in terms of resources that we can spare.

4.1   Choosing cipher algorithms

Until recently, even the smallest, lightweight ciphers like Blowfish still required minimal block or key sizes of 128 bits and a substantial number of processor cycles to execute. Since the introduction of the Speck lightweight cipher block sizes down to 32 bits are possible and the algorithms are well suited to be handled by limited performance microcontrollers. By itself, all of these security algorithms do not protect from a simple monitor, record, and replay attack.

Note that even the simplest cipher algorithm like a single exclusive OR (XOR) is considered unbreakable (literally safer than anything commonly used today) if the key is as big as the data and only used once. This is referred to as the one-time pad cipher.

So for a 32-bit value transferred, if we use a single one-time 32-bit key combined with a single XOR, we already have an encryption stronger than any other cryptography method in use today.

To a certain extent (depending on how much communication overhead is used and how often), the CANcrypt system introduced in this book allows providing such an individual key. However, "the best protection available" is hardly required for CAN communication. So even if the same or just a slightly different key is used a few times, the protection would still be adequate.

A general rule for many security systems is that the more often you use a key, the more data a possible attacker has available to analyze what is happening. If keys are only temporary and never used again, the attacker has little to work with.

The configurable CANcrypt system uses the following security features:

- Configurable and customizable algorithms for
    o   Generation and update of one-time pad
    o   One-time pad generation based on Speck or AES-128 (Advanced Encryption Standard)
    o   Checksum / hash calculation for authentication
    o   Encryption and decryption
- Secure message size is 128 bits (two CAN messages)
    o   Supporting 128-bit based algorithms such as AES-128
- All keys are synchronous and shared among two or multiple communication partners
    o   Current key used is the dynamic key, which changes after every use (used to generate pseudo one-time pad).
    o   Permanent keys (hard coded or stored in non-volatile memory) are used for initialization of the dynamic key
    o   Support of a key hierarchy (manufacturer, integrator, owner)

4.2   Elementary function: bit generation

The elementary functionality that CANcrypt provides is the generation of a bit that is known to two communication partners but not visible to anyone else. This can be a random bit, or one of the communication partners can enforce a bit. Two devices can use the bit to secretly exchange (or generate) a key. As this operation can occur at any time during operation, keys can become dynamic: new bits are introduced or added to the shared key continuously during the operation.

With this base functionality, we can pair two devices, and if the main shared key is continuously updated, the encryption, decryption, and authentication algorithms may be minimal. If the key changes randomly, an attacker that has no access to the bit generation will barely have any data to work with.

In summary, for CANcrypt the focus is not on the cipher algorithm but on the key. In the default dynamic key mode, a 64-bit key (to cover the longest possible se-

cure data block of eight bytes) is used. The key is modified after every use. The
CANcrypt configuration determines how often new random bits are introduced
into this key modification.

4.2.1 The bit-generation cycle

When monitoring CAN communications on the message level, one cannot determine the device that sent an individual message because any device may transmit
any message. As an example, let us allow two devices (named dominant device
and recessive device) to transmit messages with the CAN IDs 0010h and 0011h
and data length zero. The bits transmit within a "bit select time window" that
starts with a trigger message and has a configurable length, for example 25 ms.
Each node must randomly send one of the two messages at a random time within
the time window.

At the end of the bit select time window, a trace recording of the CAN messages
exchanged will show one of the following scenarios:

1. One or two messages of CAN ID 0010h
2. One each of CAN ID 0010h and 0011h
3. One or two messages of CAN ID 0011h

Note that if two identical messages collide, they'll be visible just once on the network. If 0010h and 0011h collide, 0010h is transmitted first followed by 0011h
(basic CAN arbitration).

Let us have a closer look at case 2 – one each. If the messages are transmitted
randomly within the bit response time window, an observer has no clue as to
which device sent which message. However, the devices themselves know it! Now
a simple "if" statement can determine the random bit for both participants:

```
IF I am the configurator device
  IF I transmitted 0010h and also saw a 0011h
    common bit is 0
  ELSE IF I transmitted 0011h and also saw 0010h
    common bit is 1
  ELSE
    both used same message, no bit determined
ELSE I am a device
  IF I transmitted 0010h and also saw a 0011h
    common bit is 1
  ELSE IF I transmitted 0011h and also saw 0010h
    common bit is 0
  ELSE
    both used same message, no bit determined
```

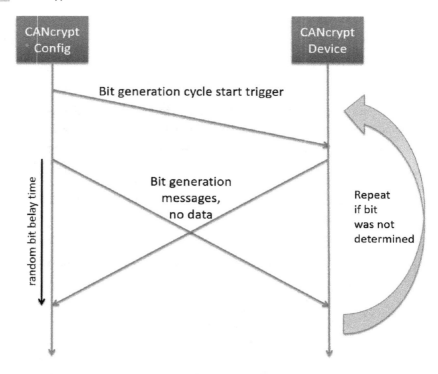

THE BIT-GENERATION CYCLE

Unfortunately we cannot use case 1 and 3, so if those happen, both nodes need to recognize it and retry – try again in the next bit select time window.

To prevent an observer from identifying individual device delays, each device should choose two good random values for each cycle. The devices should randomly pick one of the two messages (0010h or 0011h) and randomly select a delay from 0 to 2/3 of the bit select time window.

Higher-performance variations

A variation of this scheme is to not use a random delay but instead ensure that both devices directly transmit their message after the trigger message. Then both messages arbitrate the bus at the same time. In a trace recording, we will always see 0010h followed by 0011h. This scheme requires very fast reactions from the two microcontrollers using the method. From the CAN receive interrupt to the next transmit trigger, there are only a few bit times (inter frame space, seven bits), so at 1 Mbps this is just a few microseconds.

In order to minimize the chance that both devices select the same bit generation message, a variation of the scheme can use 16 or more different CAN IDs for the bit generation message. Here each device randomly selects one of the 16 messages for the bit generation. Statistically the chance that both devices select the same message is now reduced from 50% to 6%. The average duration of the complete bit-generation cycle thus shrinks drastically. The bit generation algorithm changes slightly to:

```
IF I am the configurator device
   IF I transmitted lower bit generation message
      common bit is 0
   ELSE IF I transmitted higher bit generation message
      common bit is 1
   ELSE
      both used same message, no bit determined
ELSE I am a device
   IF I transmitted lower bit generation message
      common bit is 1
   ELSE IF I transmitted higher bit generation message
      common bit is 0
   ELSE
      both used same message, no bit determined
```

4.2.2 Protection level achieved

At the logical (message) level, a bit generated or exchanged is invisible to the other communication partner. If all you can see is the CAN messages exchanged, then by monitoring CAN messages, you will not be able to determine the bit exchanged or generated.

However, an attacker who has full physical access on a signal level (oscilloscope) or on the transceiver level (connection to the microcontroller) can see which node sends which bit-select message. Nevertheless, this access only provides partial information. In CANcrypt, configurable factors, including the stored permanent key, determine how a bit is finally generated or selected.

A note in regard to random bit generation: As usual when it comes to randomness, both communication partners require a reasonably good random generator with an appropriate seed value. (If the initial seed is predictable, so is the randomness.)

4.3 Keys: generation, hierarchy and management

Each device should support multiple keys. For example, a device manufacturer might want to use a key to protect the bootloader so that only authorized and encrypted code will be loaded.

A system integrator who puts devices from multiple manufacturers into a CAN system should also be able to generate and save keys. These keys would pair the installed devices (potentially from different manufacturers) and not allow new devices to be introduced without authorization by the system integrator.

Last but not least, on the user level, it might be desirable to generate temporary keys for plug-and-play devices. A user might have authorized a particular plug-and-play device but does not want to allow additional devices or replacements without authorization.

4.3.1 Key storage in the devices

In the participating devices, permanent keys or the last session key need to be stored in non-volatile memory. Depending on the device and how it implements the storage, an intruder might try to get access to this non-volatile memory to get access to the keys.

To illustrate the vulnerability, imagine the device is Linux based and has a file system. Storing the plain keys using the file system would be simple but also an easy target. If the device also has Internet access and ever gets hacked, reading the keys from the file system becomes easy.

A device that does not use an operating system is more difficult to hack. But attackers have shown that they can load their own code if a built-in bootloader is not well protected. If the keys are just stored in a connected EEPROM, an intruder might be able to read the key.

We should make it reasonably difficult for an intruder who hacked a device to get access to the stored keys. These are some precautions we can take:

- the key storage location should not be obvious
    - if a file system is used, do not name file "keys"
    - if EEPROM is used, do not store keys at the beginning or end
    - offsets/filenames to the keys should not be constants; generate the keys dynamically as part of your initialization code
- hide the key with random data
    - if you can spare the memory, place the key within a bigger random data block
- encrypt the keys saved
    - do not store the keys as a copy;
      instead use some minimal encryption method on them

These methods do not add 100% security but raise the difficulty level for potential intruders to get easy access to the stored keys.

4.3.2 Key storage outside of the CAN system

Each time a key is generated, we need to ask ourselves if the key also needs to be stored outside the system. Potentially this creates the need to maintain a database with all keys ever generated. And that makes a very interesting target for attackers. If this option is chosen, the key copies stored need to be well protected by other security means.

Use of security "dongles"

One option for increased security could be that such keys are not stored on any PC, but only in hand-held security devices or "dongles". Only if you have physical access to one of these dongles can you make changes to the security settings of a device or system. Dongles can be based on existing CAN/CANopen handheld diagnostic tools such as CANopen Diag. A cloning function allows creating backups or copies of a dongle. The drawback is that now all keys (or a group of keys) is in one physical device, but on the plus side the keys are never stored anywhere on the Internet.

"Whoever wishes to keep a secret must hide the fact that he possesses one."

"John Golf Waggon not Heaven"

5 CANcrypt functionality

The first proof-of-concept implementations of CANcrypt were done on multiple NXP LPC17xx devices and a PC with a PEAK PCAN driver interface. Demo code is available for download at www.esacademy.com/cancrypt.

5.1 Summary

With CANcrypt, we offer a framework to handle both authentication and encryption of CAN messages. As there is some message overhead, the CANcrypt security features should be used only by a limited number of devices (the current version supports up to 15 devices) and only for selected messages (selected by CAN message ID). Depending on the chosen security level, encryption may be used not only on entire messages but also on selected bytes.

Security features are based on shared symmetric keys. There is a group key for all devices participating in the secure communication and a pairing key for secure channels between two devices. The secure pairing channel has a higher security level for use in system configuration or especially sensitive point-to-point connections such as bootloader communication.

5.1.1 Pairing

The CANcrypt pairing mode connects a CANcrypt configurator with a CANcrypt device and provides a secure communication channel supporting both authentication and encryption.

Secure messages are transmitted in pairs, first a preamble message that contains security configuration details and a signature followed by the message with the data.

The dynamic pairing key used between paired devices is continuously updated by introducing new bits generated as described in section (4.2.1 "The bit-generation cycle"). The update frequency is configurable.

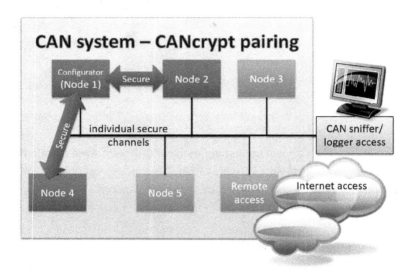

SECURE CHANNELS IN A CAN SYSTEM

5.1.2   Grouping

The CANcrypt grouping mode establishes a group of secure devices. In this mode, every device produces a secure heartbeat. The dynamic grouping key is updated based on random values in the heartbeats. No other messages use security features.

All grouped devices monitor the network for manipulations (injections, collisions in the data field) and stop producing the secure heartbeat on detecting such a manipulation.

Receiving a secure heartbeat indicates that all previous messages from the transmitting device are authentic – otherwise the device would not have produced the secure heartbeat.

Note: due to application specific delays in drivers and buffers it might be necessary to wait for two following secure heartbeats before considering a message authenticated.

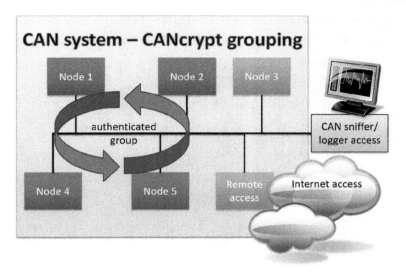

AUTHENTICATED GROUPING IN A CAN SYSTEM

5.2   Basic functionality

In this section, we outline the basic functionality provided by CANcrypt. This includes generation and updates of keys, generation of the one-time pad, and the generation and evaluation of the secure message pair.

5.2.1   Key management and key hierarchy

Security systems require keys. Security keys require management. Who keeps a copy of which key where? Does a manufacturer need to keep a copy of each individual key of every product ever produced? Which keys does a system builder or integrator need access to?

To support multiple keys at different security levels (for example for the manufacturer, system integrator, and owner of a system), CANcrypt implements a key hierarchy of up to six keys. Each of these keys has a key ID, and the higher the value for a key ID, the higher the security level.

Keys can never be read from a CANcrypt device. They can only be erased or newly generated. To erase a key, a configurator must establish a direct secure connection (active pairing) to a single device based on one of the stored keys. Once the devices are paired, the configurator can erase keys of the same or lower hierarchy level only.

In summary: once a key is generated and saved, it can only be erased and re-generated if paired based on a key of the same or higher security level.

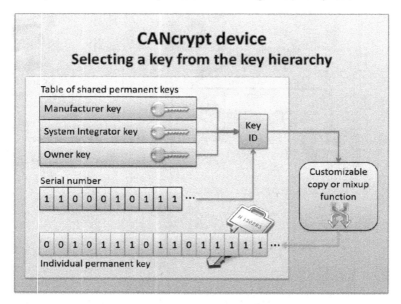

KEY SELECTION FROM KEY HIERARCHY

The pairing process requires one permanent key and may also involve an optional serial number as illustrated in the figure above, "Key selection from key hierarchy". This method allows a manufacturer to use the same base key in multiple devices. As pairing (establishing a secure channel) may also involve the serial number, a service or maintenance login could still be device specific.

5.2.2 Updating the shared dynamic keys

The dynamic key gets continuously updated following a fixed time scheme. Depending on the configuration, typical update cycle times are 500 ms, 1 s, or 2 s.

For a single pair of devices, a single new bit is generated randomly, imitated by the configurator. With multiple devices, the secure heartbeat is used to introduce new random values to by all participants.

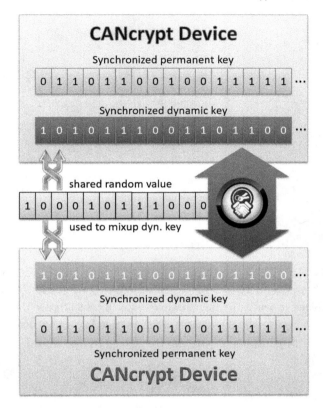

DYNAMIC KEY GENERATION WITH SHARED RANDOM NUMBERS

As part of the secure heartbeat, all participating (grouped) devices exchange encrypted random numbers. These shared random numbers are used to generate a new synchronized shared key as illustrated in the figure above. Up to 15 devices can actively participate in this mechanism.

This dynamic key is re-generated with every secure heartbeat cycle.

In paired mode (only two devices involved), the random-bit-generation cycle is used to introduce new bits to the shared dynamic key.

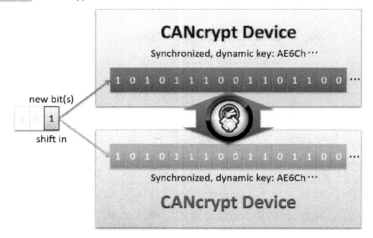

ADDING A NEW BIT TO THE DYNAMIC KEY

The new bit or bits get shifted into the dynamic key (shift right). This is done in parallel by both paired devices as illustrated in the figure above, "Adding a new bit to the dynamic key". The figure below, "New bit is shifted in", shows the new dynamic key now used by the devices. This updated key is now used for future pseudo one-time pad generations until a new bit gets introduced.

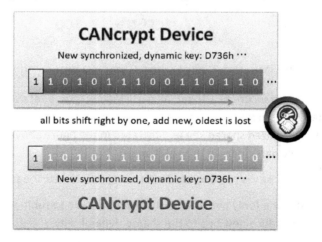

NEW BIT IS SHIFTED IN

Synchronization challenge

Even if the key update is executed by all CANcrypt devices in parallel, a secure message might still be received using the previous key. Therefore all devices must keep a copy of the previous dynamic key to decrypt and authorize messages that still use the previous key until the key update has been executed by all nodes.

5.2.3 One-time pad generation

Besides the shared dynamic key, devices also share the permanent key and a message counter (not secret) as illustrated in the figure below, "Shared parameters for pseudo one-time pad generation". The message counter is part of every secure message pair and is transmitted with the preamble message.

The dynamic one-time pad is regenerated with each transmit or receive of a secured message. The value is based on the current dynamic key, but the bits are rotated and mixed depending on a combination of the current transmit message counter and the permanent key. This method ensures that the dynamic one-time pad's bits experience a significant change between each use. Each device needs to maintain two message counters, one for transmit and one for receive, to be able to create the corresponding dynamic one-time pad.

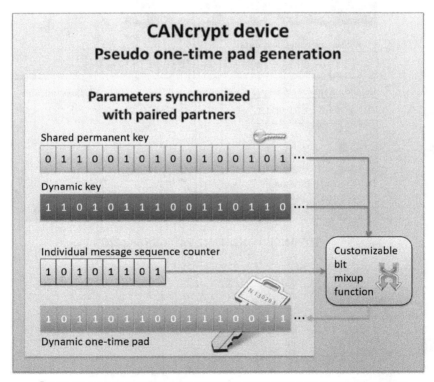

SHARED PARAMETERS FOR PSEUDO ONE-TIME PAD GENERATION

In an advanced custom version of CANcrypt additional inputs can be used for the generation of the one-time pad. This can involve decrypted data from previously received messages, for example from the secure heartbeats. Instead of light-weight Speck bit mixup function, the more advanced AES-128 or AES-256 algorithm can be used to create the one-time pad.

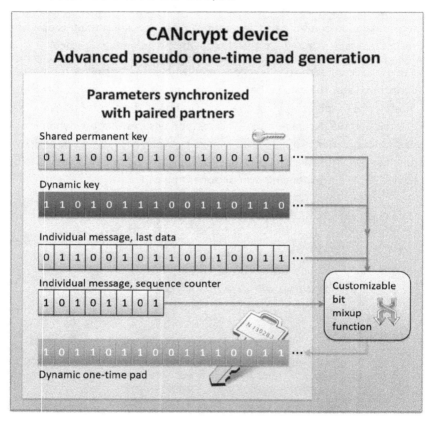

ADVANCED ONE-TIME PAD GENERATION

5.2.4 Generating an initializer for the CANcrypt checksum

CANcrypt uses checksums for the secure message table (containing configuration data), the secure heartbeat and secure messages. The checksum for the message table is calculated with an initializer of FFFFh.

As the other checksums only cover a limited number of bytes, they must not use a simple initializer like zero or FFFFh. The default CANcrypt configuration is that the initializer for these checksums is taken from a combination of the permanent key and the dynamic one-time pad. This ensures that the initializer varies from message to message.

For advanced use, the function generating the initializer can be customized, or a completely different scheme like a security hash function or a safety protocol usable checksum may be implemented.

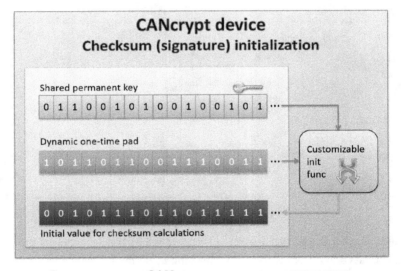

GENERATION OF CANCRYPT CHECKSUM INITIALIZER

5.2.5 Transmitting the CANcrypt Secure Heartbeat

The secure heartbeat consists of five bytes. The first byte is the heartbeat status byte indicating if the device is actively paired and the communication is still authenticated. The following three bytes are random numbers. The last byte is a checksum of the status byte and the three random bytes.

Before transmitting, the last four bytes (random number and checksum) are encrypted based on the current dynamic key.

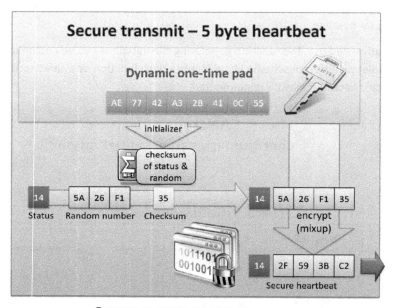

SECURE HEARTBEAT GENERATION

The timing for the secure heartbeat is controlled by the detection of other secure heartbeats and an event and an inhibit time as defined in CANopen.

On receiving a secure heartbeat, a CANcrypt device participates in the heartbeat cycle by transmitting its own secure heartbeat and resetting its internal timer. Any device detecting an expiration of the event time starts the next secure heartbeat cycle by transmitting its own secure heartbeat.

Any device may start the next heartbeat cycle earlier. For example, a device might want to have a received message authenticated as fast as possible. However, the device must wait until the inhibit time has passed before initiating a new cycle. This delay ensures that the CAN bus is not flooded with heartbeat cycles.

5.2.6 Receiving a CANcrypt Secure Heartbeat

On receiving a secure heartbeat, a device first decrypts the last four bytes based on the current shared dynamic key. Then the device calculates the checksum of the status byte and the random bytes. If the calculated checksum matches the transmitted checksum, the heartbeat is considered authenticated.

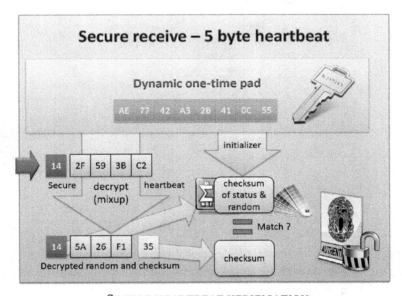

SECURE HEARTBEAT VERIFICATION

On receiving a secure heartbeat, a device determines if the local inhibit time has expired – if the time since the last transmission is greater than the inhibit time. If so, the device transmits its own next secure heartbeat.

5.2.7 Transmitting a secured message

The secure transmit handler checks if a message to be transmitted is in the global configuration list for secure messages.

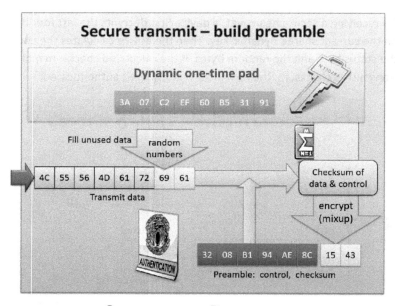

GENERATING THE PREAMBLE

If the message to be transmitted is in the list of secure messages, a preamble message is generated. The message contains control bytes including the CAN message ID to follow and the current transmit message counter.

Then the 16-bit signature is generated by calculating a checksum and encrypting it (the method is configurable; the default is an exclusive OR). Encryption happens based on the dynamic pseudo one-time pad.

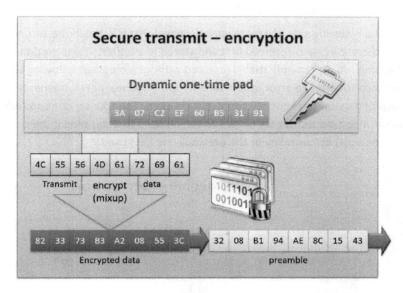

ENCRYPTING TRANSMIT DATA

If encryption for the data is used, then the data bytes requiring it are encrypted (configurable, default is exclusive OR), also based on the pseudo one-time pad. Both messages are transmitted back-to-back on the CAN system.

5.2.8 Receiving a secured message

On the receiving side, if a message is received that is listed in the secure message list, the preamble is received first and stored in a buffer. The reception starts a 10 ms timeout. A preamble that is received without a message following within 10 ms is considered an error, and a security error counter gets incremented. The included message sequence counter is checked. The counter contains information about the dynamic update cycle and can be used to determine if the current (latest, newest) dynamic key or the previous one gets used.

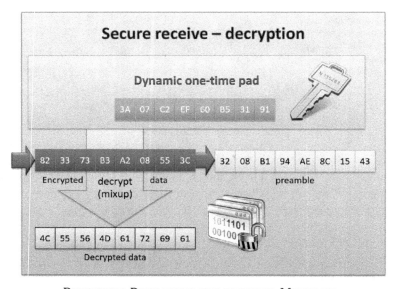

RECEIVING PREAMBLE AND SECURED MESSAGE

Once the secured message is received, the local pseudo one-time pad is generated using the message sequence counter from the preamble. If parts of the message are encrypted, they are now decrypted.

Next, the signature needs to be verified. To do that, the checksum is rebuilt on the preamble controls and the message data. The signature received with the preamble is decrypted and the two are compared. If they match, the message is considered authenticated.

A secure message received is passed on to the application or protocols above the CANcrypt handler only if authentication was successful. Otherwise the message does not get passed on and is "invisible" to a device.

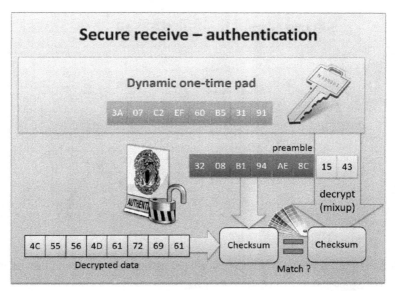

AUTHENTICATION BY RECEIVER

5.2.9 Multi message stream with epilog message

There might be situations where larger data blocks are transmitted with back-to-back messages. To optimize these transfers (by not requiring a preamble with every message), CANcrypt supports message sequence handling for up to eight messages.

If the control/request byte in the preamble specifies that multiple messages are part of the sequence, the signature of the preamble is not used. Instead an epilog message is inserted at the end of the sequence. The format of the epilog is identical to the preamble and contains the signature for all messages transferred with the sequence. See section 7.3.2 Secure stream with preamble and epilog for details about the message stream.

5.2.10 Transmitting an advanced secured message

In advanced mode, the generation of a secure message can be based on AES-128 encryption. As unused bytes are filled with random data, the total message size (consisting of the preamble and data message) is exactly 128 bits. The following figures illustrate the processes involved for encryption and decryption.

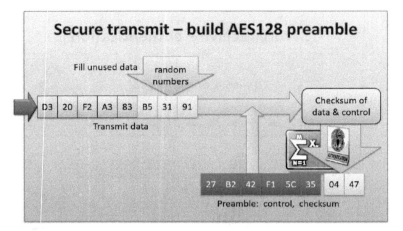

GENERATING THE PREAMBLE – AES-128 VERSION

Unused bytes in the data message are filled with random values. The preamble and data message are each eight bytes.

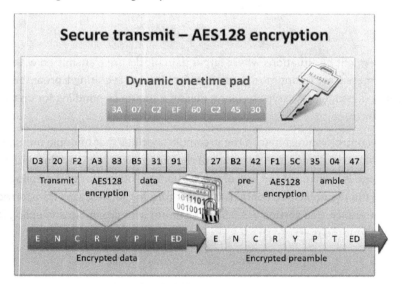

ENCRYPTING TRANSMIT DATA – AES-128 VERSION

Then both the preamble and data message are AES-128 encrypted using the current dynamic one-time pad.

5.2.11   Receiving an advanced secured message

In advanced mode, reception of a secure message requires decrypting the en-crypted values.

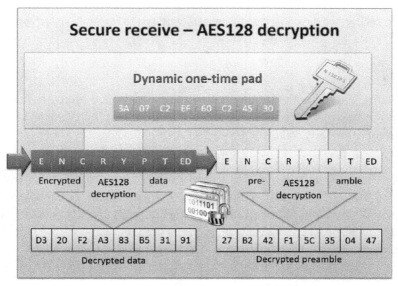

DECRYPTING RECEIVE DATA – AES-128 VERSION

Both the preamble and the secure data message are received and decrypted using AES-128 based on the current dynamic key.

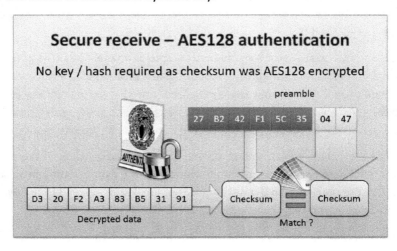

SECURE MESSAGE AUTHENTICATION – AES-128 VERSION

The receiver regenerates the checksum and compares it with the received checksum. If the checksums match, the message is considered authenticated.

5.3   Common CANcrypt parameters

In this section we describe the parameters required to maintain CANcrypt.

5.3.1   Device numbering and addressing

Address

In all CANcrypt request or command messages, a 4-bit value addresses the target CANcrypt device. A value of zero broadcasts to all devices (for example, used by the identify request). Values 1–14 are for CANcrypt devices 1–14. Address 15 is reserved for the CANcrypt configurator.

To simplify code optimizations, the addresses should be assigned incrementally starting with 1. In the CANcrypt implementation, a parameter can be set to the "highest address used". If this is set to a value below 14, CANcrypt devices using an address higher than that value must not be used (besides the CANcrypt configurator).

5.3.2   The Keys

CANcrypt supports a number of permanent keys. This allows having multiple keys per device, such as a manufacturer key for bootloader access, a system key (created upon first startup of a CAN system), or further application-specific keys or session-limited keys. For any key stored in non-volatile memory, the size is in the range 128 –1024 bits.

The main keys used are the dynamic key and the permanent key. The permanent key is the non-volatile stored key used for the initialization of the current secure communication. The dynamic key is initialized from that permanent key (a direct copy or generated using a common mixup function) and continuously modified either based on the random bit-select cycles or via the bit-update request.

The last session key can store the dynamic key over a power cycle. If there is a proper shut down procedure before power down, the dynamic key can be saved as the last session key. On the next power up, the key is reloaded to the dynamic key, drastically shortening the initialization phase.

To globally identify the keys, CANcrypt uses 8-bit Key ID and Key length parameters. These values are used as described below.

Key ID

The Key ID is divided into a 3-bit major value and a 5-bit minor value.

The major value specifies one of eight key types and directly implements a key hierarchy. Higher values have a higher authority. The key erase command can be used only on keys that have the same or lower major value as the key currently in use.

The minor value plus specifies 32-bit segments within the key.

The key length value determines, if a key is used by itself without modifications or gets combined (mixed up) with the local serial number.

The values are mapped to UNSIGNED8 values. The major part uses the three most significant bits, and the minor part uses the five least significant bits.

Default use	Memory	Key ID major	Key ID minor	Length (bit)
Reserved		7		
Manufacturer key	NVOL	6	0–31	128–1024
System Integration key	NVOL	5	0–31	128–1024
Owner key	NVOL	4	0–31	128–1024
User key	NVOL	3	0–31	128–1024
Last group session key	NVOL	2	0–15	128–512
Dynamic pair session key	RAM	1	0–15	128–512
Dynamic group session key	RAM	0	0–15	128–512

THE KEY HIERARCHY

Key length

The Key Length is of type UNSIGNED8. To support a wide variety of key lengths with 8-bit encoding, the highest bit determines if the size is specified in bits or in other units as shown in the table below (Key Length Values Supported by CANcrypt).

Value	Interpretation
00h	Reserved
01h–20h	Key length in bits, 1–32
21h–7Fh	Reserved
80h	Single bit of dynamic key
81h–A0h	Key length in multiples of 32 bits, 1–32 (32–1024 bits)
A1h–C0h	As above, but key is combined with serial number
C1h–FFh	Custom, manufacturer specific sizes

KEY LENGTH VALUES SUPPORTED BY CANCRYPT

5.3.3 Status

This section describes the status information that must be provided by all partici-pating CANcrypt communication partners.

Status

The CANcrypt status byte provides the following information and is the same for both the CANcrypt configurator and devices:

- Bits 0–1:   Pairing status
              0: not paired
              1: pairing in progress
              2: paired
              3: pairing error
- Bits 2–3:   Grouping status
              0: not grouped
              1: grouping in progress
              2: grouped, secure heartbeat enabled
              3: grouping error
- Bits 4–5:   Result of last command or request
              0: unknown
              1: success
              2: ignored
              3: failure
- Bit 6:   Reserved
- Bit 7:   Key generation in progress
              When set, this device is participating in key generation

5.3.4 Controls

This section describes the control commands and requests available to the CAN-crypt configurator and devices.

Request and commands

The 4-bit request value is used in most CANcrypt protocols.

Message	Type	Consumer Address	Request
Abort	event, response	1–15	0
Acknowledge	response	1–15	1
Alert	event	0	2
Identify	event	0	3
Pairing/Grouping	request, response	1–15	4
Unpairing	request, response	1–15	5
Flip bits in key	request	1–15	6
Bit or key generation	request, response	1–15	7
Bit generation trigger	request	0	8
Secure heartbeat	Event	0	9
Generic data read	secure exchange	1–15	10
Generic data write	secure exchange	1–15	11
Generic data ID	secure exchange	1–15	12
Extended Identify	request, response	1-15	13
Preamble, Epilog	event	0	14
Save last session key	event	0	15

REQUESTS USED BY CANCRYPT DEVICES AND CONFIGURATOR

The requests and commands in the table "Requests used by CANcrypt devices and configurator" are used by both devices and the configurator in the same manner.

There is one exception: the identify request and response. When used by the configurator, these requests have extra parameters. For details, see the protocol definitions in Chapter 6.

5.3.5 Methods

CANcrypt supports a variety of algorithms and features. The parameters selecting these are listed below. For more details about the specific algorithms used, see chapter 6, CANcrypt customizable functions.

Method

The 4-bit method parameter selects the base algorithm used to generate the random bit and specifies a security method.

- Bits 0–1:   Security functionality
                    0: Basic security
                    1: Regular security
                    2: Advanced Security
                    3: Custom security
- Bit 2:   Bit generation method, set to 1 for random delay,
                    otherwise direct, immediate reply to trigger message.
- Bit 3:   Number of bit generation messages used. When set, 16
                    bit generation messages are used, else 2.

The security settings influence the bit-generation cycle, authentication, and en-cryption.

Bit generation:
After each bit-generation cycle, the customizable function UpdateBit() is called and can flip the bit generated, for example depending on the permanent key. This increases security for cases where an intruder has physical access to the CAN system as the intruder cannot easily determine when a new bit generated is 0 or 1. In addition, bit stuffing is used.. This ensures that the Mixup() function used for authentication and encryption does not use a value with all the same bits.

Authentication:
The signature used for messages is 16 bits. The signature is generated by the combination of a checksum that is encrypted using a bit mixup of the current dynamic key and the message counter. In basic mode, the checksum is calculated in Fletcher style with the initialization generated from the permanent or dynamic key. In regular mode or higher a 16bit CRC checksum is used. In advanced mode AES-128 is used for encryption/decryption.

Encryption:

The encryption is based on a mixup of the current dynamic key.

5.3.6 Functionality

Individual CANcrypt functionality may be enabled or disabled.

Functionality

If a corresponding bit is set, the functionality is enabled

- Bit 0:   authentication used
- Bit 1:   encryption used
- Bits 2–3:   reserved

5.3.7 Timings

CANcrypt uses various timings and timeouts. To minimize the number of definitions, specific values are defined as a group.

Timeout

The 4-bit timeout value defines the timing and timeout options CANcrypt uses:

- Bits 0–1:   timing used
                     0: fast
                     1: medium
                     2: slow
                     3: custom timing
- Bits 2–3:   reserved

Values 0–2 activate the defined timings in the table below, Timeouts Used by CANCrypt). Value 3 selects custom, manufacturer-specific timings.

CANcyrpt message timeout:

If a CANcrypt message contains a request, requiring a response, then the transmitter uses this timeout to wait for an response from the device addressed. If no response is received within this time, the transmitter internally marks the addressed device as not present.

Secure message timeout:

Every secure message combination using a preamble and one or multiple following data messages have to transmit the messages back to back on the network. On the receiving side the data message is only considered to be received in time, if the time since reception of the preamble does not exceed this timeout.

Timeouts		Fast	Medium	Slow
CANcrypt message timeout (request to response)		100 ms	200 ms	400 ms
Secure message timeout (preamble to message)		25 ms	50 ms	100 ms
Secure heartbeat event time (slowest repetition)		250 ms	500 ms	1000 ms
Secure heartbeat event timeout		500 ms	1 s	2 s
Secure heartbeat inhibit time (fastest repetition)		50 ms	100 ms	250 ms
Secure heartbeat cycle timeout		75 ms	150 ms	333 ms
Bit select cycle time for random delay method		25 ms	50 ms	100 ms
Bit select cycle time for direct response method with no delay		10 ms	25 ms	50 ms
Bit select cycle random delay window		0–16 ms	0–32 ms	0–64 ms

DEFAULT TIMEOUTS USED BY CANCRYPT

Secure heartbeat event time and timeout:
The longest possible duration between two secure heartbeat cycles is defined by the event time. A device is considered unsecure or missing if the time since the last secure heartbeat transmission exceeds the timeout.

Secure heartbeat inhibit time and cycle timeout:
The shortest possible duration between two secure heartbeat cycles is defined by the inhibit time. All devices may start a new secure heartbeat cycle at any time, as long as they ensure that the inhibit time is met. If a secure heartbeat cycle started, than all active devices must join the cycle with their own secure heartbeat within the cycle timeout. A device not participating in time is considered unsecure or missing.

Bit select cycle time and delay window:
The key- or bit-generation cycle time is a fixed value, the CANcrypt system tries to determine one bit per cycle. If the method with delays is used (each participant

transmits their claim message randomly within a time window), then the maximum value for this delay is defined.

5.3.8   CANcrypt error counter

CAN uses transmit and receive error counters to determine the "health" of an individual CAN controller. When errors occur, the timers are incremented by a number greater than 1. However, the timers are also decremented when communication works fine. As a result, occasional errors are ignored. But if the counters keep increasing and hit limits, the CAN controller goes "passive" or eventually "bus off," which is a complete disconnection of the CAN controller from the network.

Event	Counter change
Successful reception of a secure message or secure heartbeat (no timeout, successful authentication)	If counter > 0, decrement
Secure Heartbeat failure (timeout or authentication failure)	Set counter to 128
Intruder alert (injected message with harmful data detected)	Set counter to 128
Intruder alert (injected message with harmless data detected)	Increment counter by 63
Secure message authentication failure	Increment counter by 63
Other error in secure message (preamble timeout, receive without preamble)	Increment counter by 31
Repetitive request from same device to re-initialize the dynamic key	Increment counter by 31
Any other alert event, CAN errors	Increment counter by 15

CANCRYPT ERROR COUNTER CHANGES

In the same manner, CANcrypt uses an UNSIGNED8 error counter to determine the health of the CANcrypt connection. The table below (CANCrypt Error Counter Changes) shows which events influence the error counter. Once the error counter reaches 128 or higher, the CANcrypt device unpairs, or disconnects itself, from secure communication and uses the unpair protocol/status to inform all other devices.

Once a device unpairs itself, it should not be allowed to immediately participate in a re-pairing process. A generous timeout should be required before a retry of the re-pairing starts. If the unpairing is a result of an attack, the intruder may try brute-force methods to participate in pairing processes. To slow such attacks, every failed pairing attempt should cause a delay of increasing seconds.

Even if your application requires constant operation, keep in mind that once we reach the state of unpairing, we are either under attack or something went seriously wrong (device disconnected or powered down).

5.4   CANcrypt secure message table

Datatype	Name	Use
UNSIGNED32	CAN ID	CAN ID of the secure message. Set bit 30 to indicate that a 29-bit CAN ID is used.
UNSIGNED8	First encrypted byte	If using encryption, the first byte to which encryption is applied (starting at zero).
UNSIGNED8	Number of encrypted bytes	If using encryption, the number of encrypted bytes.
UNSIGNED4	Functionality	CANcrpyt functionality used for this message.
UNSIGNED4	Method	CANcrypt methods used for this message.
UNSIGNED4	Producer	CANcrypt address of the device producing this message (1–14).
UNSIGNED4	Reserved	

ENTRY IN THE SECURE MESSAGE LIST TABLE

The last entry of the table is different, see below.

Datatype	Name	Use
UNSIGNED32	End of table	Set to FFFF FFFFh to mark the end of the table.
UNSIGNED16	Reserved	Set to FFFFh
UNSIGNED16	Checksum	Checksum covering all table entries without the End of Table entry.

LAST ENTRY IN THE SECURE MESSAGE LIST TABLE

Note that for optimization individual devices may only store those elements of the table that they require. If a message is not used by a local device, its details do not need to be known by the device.

Each element in the table is 8 bytes and provides details about a secure message handled by CANcrypt. The last entry in the table must have a CAN ID of FFFF FFFFh to indicate the end of the table. The last record also uses a 16-bit checksum for the entire table.

If a high security level is desired, the configuration options for the secure message table should be limited. An intruder with access to this level (being able to edit the table) could reconfigure a device to listen for different messages other than those originally intended.

A typical use case would be that the table initially gets set by the configurator using secure communication as defined in 7.3.3 Secure generic data object access. Once the table is programmed, modifications should only be possible by a configurator with the same permanent key access used to program the table in the first place.

The checksum method used shall be the highest level method supported by a device. If a device supports only the regular security method, the checksum method of that level is used. The checksum initializer for this checksum shall be FFFFh.

For better optimization, each device uses two local tables, one for secure messages received by this device and one for secure messages transmitted.

5.4.1 Pairing and grouping implementation note

As the mechanisms used to produce and consume secure messages are the same for a paired and a grouped communication, the tables and other resources required may be shared for both paired and grouped communication.

However, when resources are shared, secure communication cannot be used at the same time by a device that is both grouped and paired. If the application requires that secure communication is possible at the same time for both paired and grouped devices, then the keys and tables need to be duplicated, one set for the paired communication and one set for the grouped communication.

5.5 Use case example: secure push button

Any CANcrypt device that can permanently store keys in non-volatile memory needs theses key generated and stored at some point. Let us examine the example of a device with a manual push button. This button later activates a function that is considered secure, like unlocking some door.

5.5.1 Manufacturer

At the end of the production line, a manufacturer has the option to generate and store a manufacturer key (bootloader, manufacturer reset). Note that in some cases, a manufacturer may choose to have this key hard coded in firmware.

It is up to the manufacturer whether to install different manufacturer keys for each device or shared keys, like one per product variation. For example, all products of the same order code and revision number could share a key.

Note that CANcrypt optionally allows combining the permanent key with a serial number so even with a shared key, individual keys could still differ depending on serial number.

For this example, let us say that the push button device has a serial number and a shared manufacturer reset key that is hard coded in firmware – pairing on this level allows erasing all other security keys of a lesser security level.

5.5.2 System integrator: reset key

A system builder installs multiple devices with CANcrypt support from multiple vendors in a CAN system. For this example, the devices include the push button from the previous section but also the locker to a door. Using a CANcrypt configurator (PC or handheld), the system builder creates a common "system builder" reset key for all devices. This allows the system builder to reset (erase all other keys of a lower security level) these devices if needed for maintenance.

5.5.3 System integrator: grouping

Once all devices are installed, the configurator loads the CANcrypt configuration for each device including device ID and secure message tables.

Finally the system builder performs a grouping of all CANcrypt devices in the system using a CANcrypt configurator. This key does not get stored by the configurator. Only the CANcrypt devices save this key. Finally, an initial "last session key" is generated and saved by all devices.

Now all CANcrypt devices in the system are grouped, and the key used is stored only by the devices in the CAN system, not outside.

5.5.4 System start

All CANcrypt devices use the startup protocol to start the system. If all previously present devices are still present and can use their last saved session key, system start can be completed in as little as 100 ms.

If one or more devices report that they cannot use the last session key, they will try to use the next available key in the key hierarchy to activate the system. In these cases, the dynamic key is initialized based on the key selected and the random numbers provided by all devices.

5.5.5 Support case: replace the button

If in the system above, the push button needs to be replaced, the system builder first needs to erase all grouping keys. Using the CANcrypt configurator, the system builder uses the system builder reset key to erase all grouping keys.

The new push button gets the system builder reset key assigned, and a new configuration and grouping cycle is started.

If the broken push button can no longer be accessed by the system builder, the manufacturer has two options to help the system builder reset the device to manufacturer defaults:

a) have the device returned to the manufacturer and use the manufacturer key to access the device
b) Send the "trusted" system builder the individual key (combined with serial number) to access the device

"Those who cannot change their minds cannot change anything."

"Roger Shrew Bandage"

6 CANcrypt customizable functions

CANcrypt uses a value in the range of 0–3 to select one of four security function levels as shown in the table below (selecting CANcrypt methods and algorithms).

Name	Value	Description
Basic	0	Minimal security level, requires minimal computational resources, usable on most microcontrollers. Cryptographic method used is the 64-bit Speck Cipher with limited rounds (7). Only protects from accidental misuse and simple record and replay scenarios.
Regular	1	Default security level, adequate for all applications without specific security requirements, suited for 32-bit microcontrollers. Cryptographic method used is the 64-bit Speck Cipher with full rounds (27)
Advanced	2	Highest security level, potentially suitable to also provide safety functionality. Uses AES-128.
Custom	3	Allows customization of all security relevant functions.

SELECTING CANCRYPT DEFAULT METHODS AND ALGORITHMS

All elementary CANcrypt functions that actively influence the security level are located in the *CANcrypt_userfct.h* module. System developers can select either one of the default settings or use their own customized configuration. In this chapter we show the provided functions, for the default implementation see Appendix B.

```
/*******************************************************************
MODULE:     CANcrypt_userfct.h, customizable functions
CONTAINS:   Customizable security functions of CANcryot
AUTHOR:     2017 Embedded Systems Academy, GmbH
HOME:       www.esacademy.com/cancrypt

   Licensed under the Apache License, Version 2.0 (the "License");
   you may not use this file except in compliance with the License.
   You may obtain a copy of the License at
   www.apache.org/licenses/LICENSE-2.0

   Unless required by applicable law or agreed to in writing, software
   distributed under the License is distributed on an "AS IS" BASIS,
   WITHOUT WARRANTIES OR CONDITIONS OF ANY KIND, either express or
   implied. See the License for the specific language governing
   permissions and limitations under the License.

VERSION:    0.10, 19-JAN-2017
********************************************************************/

#ifndef _CANCRYPT_USERFCT_H
#define _CANCRYPT_USERFCT_H

#include "CANcrypt_types.h"

#if (Cc_FUNCTIONALITY == Cc_SECFCT_ADVANCED)
// Not part of CANcrypt, AES implementation
extern void AESxxx_Encryption(
   UNSIGNED32 *data,      // input 1: data to encrypt
   UNSIGNED32 *key,       // input 2: key to use
   UNSIGNED32 *coded      // output: encrypted data
   );
extern void AESxxx_Decryption(
   UNSIGNED32 *data,      // input 1: data to decrypt
   UNSIGNED32 *key,       // input 2: key to use
   UNSIGNED32 *coded      // output: decrypted data
   );
#endif
```

The advanced method uses AES encryption. Code to implement AES is not part of CANcrypt. Some microcontroller systems have dedicated hardware engines for AES encoding and decoding. If you need AES implemented in software, there are several open source implementations available, for example on the "GitHub" or "SourceForge" code sharing platforms.

6.1   Collect random numbers

Both pairing and grouping functions collect random data from the participating devices to initialize the secure communication. For the initial key generation, CANcrypt requires an array of random numbers that is as long as the current key length. This function takes the collected initial random numbers and expands them to fill the required array.

```
/*****************************************************************
BOOK:    Section 6.1 "Collect random numbers"
DOES:    This function expands an array with a limited number of
         random bytes to an array of random bytes with the length
         of the current dynamic key.
RETURNS: nothing
*****************************************************************/
void Ccuser_ExpandRandom(
  UNSIGNED32 *pkey,      // key input, length Cc_PERMKEY_LEN_BITS
  UNSIGNED32 *pdest,     // destination: array with length of dyn. key
  UNSIGNED32 *psrc       // array with zeros and random numbers (3*15)
);
```

6.2   Bit mixup

The bit mixup function is used in basic and regular modes to generate the initial dynamic key from the permanent key and to generate the dynamic one-time pad from the current dynamic key. The default CANcrypt implementation uses a variation of the 64-bit Speck Cipher. The number of rounds executed is configurable.

```
/*****************************************************************
Macros to rotate 32bit value right or left and a single mix up round
in add-rotate-xor (ARX) style as used by Speck cipher
*****************************************************************/
#define ROR32(x,r) ( (x >> (r & 0x1F)) | (x << (32 - (r & 0x1F))) )
#define ROL32(x,l) ( (x << (l & 0x1F)) | (x >> (32 - (l & 0x1F))) )
#define MIXROUND32(a,b,k) (a=ROR32(a,8),a+=b,a^=b,b=ROL32(b,3),b^=a)


/*****************************************************************
BOOK:    Section 6.2 "Bit mixup"
DOES:    This function mixes the bits in a 64bit value by applying
         a Speck cipher. Used by key initialization functions and
         one-time pad generation.
MOTE:    Recommended number of rounds is 27
RETURNS: Value pmixed[] returns the mixed bits
*****************************************************************/
void Ccuser_Mix64(
  UNSIGNED32 *pkey,        // key input, length Cc_PERMKEY_LEN_BITS
  UNSIGNED32 pdat[2],      // data input of 64 bit
  UNSIGNED32 pmixed[2],    // mixed bits output of 64 bit
  UNSIGNED8 rounds         // number of mixing rounds to execute
);
```

6.3   Generate dynamic key

When a CANcrypt system powers up it does not yet have a shared dynamic key, only the stored permanent key. Initialization of the dynamic key depends on the connection method. In both available methods the selected permanent key is copied to the dynamic key and gets modified before its first use.

The same function is also used to update or re-generate the dynamic key in grouping mode and to generate a one-time pad from the dynamic key.

The main idea behind this is that a permanent key should never be used directly for any security functions, as that would provide an attack vector, therefore a modified copy is used.

```
/****************************************************************
BOOK:     Section 6.3 "Generate keys"
DOES:     Takes input from 2 keys and 1 factor to create a new key.
          Used to create a dynamic key from a permanent key using
          random input and a serial number.
          Used to create a one-time pad from a permanent and
          dynamic key and a counter
RETURNS:  TRUE if key initialization completed,
          FALSE if not possible due to parameters
****************************************************************/
UNSIGNED8 Ccuser_MakeKey(
  UNSIGNED32 *pin1,      // input 1: pointer to permanent key used
  UNSIGNED32 *pin2,      // input 2: pointer to 2nd input array
  UNSIGNED32 factor,     // input 3: optional, set zero if not used
                         // used for serial number, counter
  UNSIGNED32 *pout       // output: the dynamic key or one time pad
);
```

6.3.1   Pairing: Generate a random key first

When two CANcrypt devices are paired, they use the pairing messages to exchange initial random 24-bit values. The dynamic key is initialized by making a copy of the permanent key and then using the *Ccuser_MakeKey()* function with the exchanged random value. After that, at least 16 bit-generation cycles to update the key should run before the dynamic key is used.

6.3.2   Grouping: Take random values from grouping message

The messages used to initialize the grouping mode each contain a 24-bit random value. These random values from all participating devices are used to initiate the dynamic key before its initial use.

6.3.3   Generate one-time pad

The one-time pad is re-generated before every use of secure messages. An individual message counter is part of the generation, ensuring that some value changes with every use.

Note that the message counter is not used with the secure heartbeat, as here the key is updated automatically with every cycle.

6.4   Updating the dynamic shared key

One of the core features of CANcrypt is that the dynamic key is continuously updated. The methods used differ between pairing and grouping.

6.4.1   Pairing: Key update using a single bit

In pairing mode, the bit-generation cycle is executed periodically in the background. The bit generated is used to modify the shared key.

The generated bit is shifted in from the left. Additional input from the permanent key determines if the current bit is flipped or not.

```
/****************************************************************
BOOK:     Section 6.4.1 "Pairing: Key update using a single bit"
DOES:     Called to update a dynamic key by introducing a new bit.
RETURNS:  TRUE if key update completed,
          FALSE if not possible due to parameters
****************************************************************/
UNSIGNED8 Ccuser_UpdateDynKey(
  UNSIGNED8 bit,        // new bit to introduce to dynamic key
  UNSIGNED32 *ppermkey, // pointer to permanent key used
  UNSIGNED32 *pdynkey   // pointer to dynamic key
);
```

6.4.2   Grouping: Key update based on secure heartbeat

In grouping mode, a secure heartbeat is produced.

Every secure heartbeat includes three random, encrypted bytes. The secure heartbeat happens in cycles, and within each cycle, all active participants produce their heartbeats.

All decrypted random values from all devices are used to update the dynamic key.

This key update uses the *Ccuser_MakeKey()* function from section 6.3 to generate the dynamic key.

6.5   Secure Heartbeat

The secure heartbeat contains a 32-bit signature consisting of a 24-bit random value and an 8-bit checksum. The entire value is encrypted based on the current/last dynamic key used.

6.5.1   Generate secure Heartbeat value

The secure heartbeat is device specific, so the input parameters include the random value, the dynamic key and the address of the device sending the heartbeat.

```
/*******************************************************************
BOOK:    Section 6.5.1 "Generate Signature Value"
DOES:    Generates a signature for this device
RETURNS: The 32bit signature
*******************************************************************/
UNSIGNED32 Ccuser_MakeSignature(
   UNSIGNED8 address,     // device ID (1-15) of this device
   UNSIGNED32 *pdyn,      // pointer to dynamic key used
   UNSIGNED32 rnd         // new random value to introduce
);
```

6.5.2   Verify secure Heartbeat value

To verify a secure heartbeat, we need the value received, the device address from which the heartbeat was received and the dynamic key used to create it.

```
/*******************************************************************
BOOK:    Section 6.5.2 "Verify Signature Value"
DOES:    Verifies a signature received from a device
RETURNS: TRUE, if signature was verified
*******************************************************************/
UNSIGNED8 Ccuser_VerifySignature(
   UNSIGNED8 address,     // device ID (1-15) of device sending the
   UNSIGNED32 *sHB,       // signature
                          // on return, decrypted value is at location
   UNSIGNED32 *pdyn       // pointer to dynamic key used
);
```

6.6   Secure message checksum generation

The checksum used for the secure messages are used for authentication. Therefore, we recommend to use a key dependent initializer instead of the typical 0 or all Fh. In CANcrypt, checksums are always encrypted.

The default checksum calculation is in Fletcher style, see the examples provided. In this mode a 32-bit checksum is calculated and in a final step a 16-bit value is extracted from that. For advanced or custom mode we recommend to use CRC algorithms.

6.6.1   Checksum step

This function adds a single 16-bit value to the checksum. Note that parameter passed and returned is 32-bit, as initial calculation is based on 32-bit.

```
/*******************************************************************
```

```
BOOK:    Section 6.6.1 "Checksum step"
DOES:    Calculates a 16bit checksum, adding one value at the time
RETURNS: Checksum value in lowest 16bit, highest 16bit is a carry-over
*********************************************************************/
UNSIGNED32 Ccuser_ChecksumStep16(
  UNSIGNED32 last,        // initial value or last calculated value
                          // higher 16bit may include a carry-over
  UNSIGNED16 *pdat        // next 16bit value to add
);
```

6.6.2   Checksum final

This function performs the last step of the checksum calculation. Here it generates a 16-bit value from the calculated 32-bit checksum.

```
/********************************************************************
BOOK:    Section 6.6.2 "Checksum final"
DOES:    When checksum calculation is completed, merges 16bit checksum
         with 16bit carry ove rto final 16bit checksum.
RETURNS: Final checksum value
*********************************************************************/
UNSIGNED16 Ccuser_ChecksumFinal(
  UNSIGNED32 last         // last calculated checksum value
);
```

6.7   Encryption and decryption

Encryption algorithms are kept simple in CANcrypt, the security effort is placed into the dynamic keys and one-time pads.

6.7.1   Secure message encryption

When it comes to secure messaging, CANcrypt ensures that these always are made up of two CAN messages of eight bytes, providing a total data length of 128 bits. The first message is a preamble, the second the data message, with unused bytes filled with random bytes. This potentially allows 128-bit algorithms to be used if the entire message needs to be encrypted.

Per default, encryption is a single exclusive or with the current one-time pad. Only the bytes specified get encrypted.

```
/*********************************************************************
BOOK:     Section 6.7.1 "Secure message encryption"
DOES:     Encrypts a data block in a secure message
NOTE:     This version NOT optimized for 32 bit architecture
RETURNS:  TRUE if encryption completed,
          FALSE if not possible due to parameters
*********************************************************************/
UNSIGNED8 Ccuser_Encrypt(
  UNSIGNED32 *ppad,      // pointer to current one-time pad
  UNSIGNED32 *pdat,      // pointer to the data to encrypt
  UNSIGNED16 first,      // first byte to encrypt
  UNSIGNED16 bytes       // number of bytes to encrypt
);
```

6.7.2 Secure message decryption

The decryption function uses the same parameters. If the encryption method is fully symmetric, then the encrypt function can also be used for decrypt.

```
/*********************************************************************
BOOK:     Section 6.7.2 "Secure message decryption"
NOTE:     Only used if cryptographic function is not symmetric and
          decryption requires a different function then encryption
DOES:     Decrypts a data block
RETURNS:  TRUE if decryption completed,
          FALSE if not possible due to parameters
*********************************************************************/
UNSIGNED8 Ccuser_Decrypt(
  UNSIGNED32 *ppad,      // pointer to current one-time pad
  UNSIGNED32 *pdat,      // pointer to the data to encrypt
  UNSIGNED16 first,      // first byte to decrypt
  UNSIGNED16 bytes       // number of bytes to decrypt
);
```

7 CANcrypt protocol layers

In this chapter we describe the specific CAN-based protocols that implement CANcrypt. Note that this section uses the parameters defined in section 5.3 Common CANcrypt parameters.

7.1   Basic protocol elements

The basic protocol elements include selecting CAN message identifiers, protocols for events like alerts or aborts, and the random-bit-generation cycle, which is used by multiple CANcrypt protocols.

7.1.1   CAN message identifiers

We strongly recommend that the CAN message identifiers be hard coded so they cannot be reconfigured through CAN communication during operation. Otherwise, attackers could try to reconfigure the CAN ID usage. On success, they would be able to logically disconnect one of the secure devices. Doing so would be a first step in an attempt to replace a secure device with a device provided by the attacker.

For the CANcrypt configurator and devices, we need one CAN message identifier for the device's main CANcrypt command, response and status message.

To simplify implementation, the CANcrypt devices should use up to 15 consecutive identifiers. The CANcrypt configurator uses the first identifier. The identifiers should be high priority (low value). When used as preamble to a high priority message, a low-priority identifier might cause delays.

The default CANcrypt message IDs are 171h to 17Fh. The CAN message IDs 172h – 17Fh are used by the CANcrypt devices, and the CANcrypt configurator uses 171h. All devices must receive all CANcrypt message IDs and respond to requests or commands received.

The bit-generation cycle requires two CAN message IDs. Key generation or key update is a background process and may be of lower priority.

The default CANcrypt bit-generation message IDs are 6FEh and 6FFh. These are used by both the CANcrypt configurator and devices in the bit-generation cycle for generating keys or for pairing. The configuration must ensure that at any time, only one bit-generation cycle is active.

The default CAN message IDs are values that are reserved, and thus otherwise unused, by CANopen. Depending on the protocol or application, other identifiers may be used.

CAN ID	CANcrypt use
172h–17Fh	CANcrypt message of CANcrypt devices 2–15
171h	CANcrypt message of CANcrypt configurator
6F0h–6FFh	Bit-selection messages for random-bit-select cycle
6E1h–6EEh	Optional debug messages from CANcrypt devices

DEFAULT CAN IDS USED BY CANCRYPT

7.1.2   CANcrypt message common contents

The first two bytes of the CANcrypt message are identical for all requests and responses. They contain:

    address (4 bits): destination device
    access (4 bits):   message identification request, response or event
    status (8 bits):   current status byte

The address is the CANcrypt device number for the message's destination. Set to 0 for a broadcast to all devices or 1 for the configurator.

The access information identifies which request, response, or event the message contains. See section 5.3.4 for a complete list of all values and section 5.3.3 for a description of the current status byte.

The access type determines how many additional bytes follow these first two bytes and what information the bytes contain.

7.1.3 Alerts and errors

At any time, any CANcrypt device may generate an alert to signal that an error or intruder detection occurred. By itself, these signals are not secure.

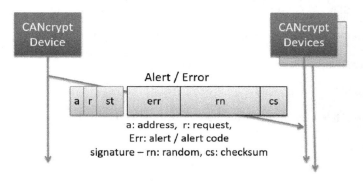

ALERT OR ERROR SIGNAL

Parameters used (8-bit unless noted otherwise):

address (4 bits): 0, broadcast
request (4 bits): 2, alert
status: current status byte
error (16 bits):   error code,
                   high byte for error / alert code
                   low byte for manufacturer specific information
sign. (32 bit)   signature, 3 random bytes, 1 byte checksum,
                   all encrypted based on dynamic key

Code	Interpretation
80h to 8Fh	Intruder alert
90h to 9Fh	Key generation error or timeout
A0h to AFh	Pairing error or timeout
B0h to BFh	Grouping or secure heartbeat error or timeout
C0h to CFh	Secure messaging error or timeout
D0h to DFh	Generic access error or timeout
E0h to EFh	Reserved
F0h to FFh	Manufacturer specific

CANCRYPT ERROR AND ALERT CODES

Alternate CANopen use

For minimal CANopen integration, the alert / error message can be embedded into the 5 manufacturer specific bytes of the CANopen emergency message. In this case the status and the error code go into the three first bytes of the five manufacturer-specific bytes. The recommended emergency codes to use are listed in the table below.

EMCY code	Description
8310h	CANcrypt alert event

7.1.4 Acknowledge or Abort

All protocols consisting of a sequence of messages may use the messages acknowledge and abort. Aborts may be used at any time by any of the involved communication partners to abort (end) the sequence. Acknowledge may only be used as specified by the individual sequence. By itself, these signals are not secure.

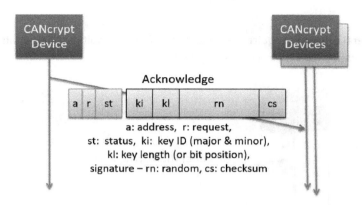

ACKNOWLEDGE OR ABORT

Parameters used (8-bit unless noted otherwise):

address (4 bits):	1–15, device address
request (4 bits):	0, abort or 1, acknowledge
status:	current status byte
key ID:	if key is involved, ID of key
key length:	if key is involved, length of key
sign. (32 bit)	signature, 3 random bytes, 1 byte checksum, all encrypted based on dynamic key

Alternate CANopen use

For minimal CANopen integration, the acknowledge / abort message can be embedded into the five manufacturer specific bytes of the CANopen emergency message. In this case the status and the key ID and length go into the three first bytes of the five manufacturer-specific bytes. The recommended emergency codes to use are listed in the table below.

EMCY code	Description
0090h	Generic CANcrypt acknowledge
8300h	Generic CANcrypt sequence abort

7.1.5 Sub-protocol for bit-generation

Several protocols require the generation of one or multiple shared bits. Each bit-generation uses the cycle outlined in this section.

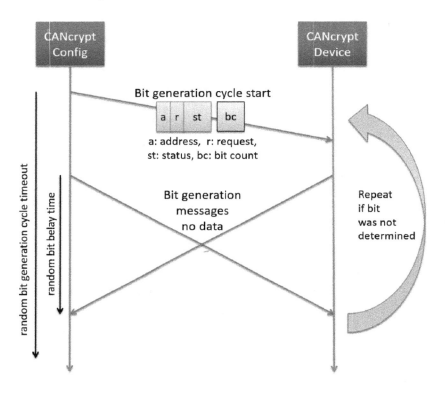

THE BIT-GENERATION CYCLE PROTOCOL

Parameters used (8-bit unless noted otherwise):

address (4 bits):	1–15, device address
request (4 bits):	4, pairing
	7, bit or key generation
status:	current status byte
bit count:	counting down from 31 to 1 or
	80h if this is a single bit for the dynamic key

Each cycle starts with a trigger message from the CANcrypt device initiating the bit-generation. Depending on the mode used, each device transmits its chosen bit-select message immediately or with a random delay. If a bit was not determined (both partners used an identical message), the cycle is repeated

The following flow charts illustrate the processes executed internally in the CANcrypt Configurator and CANcrypt device during the bit-generation cycle.

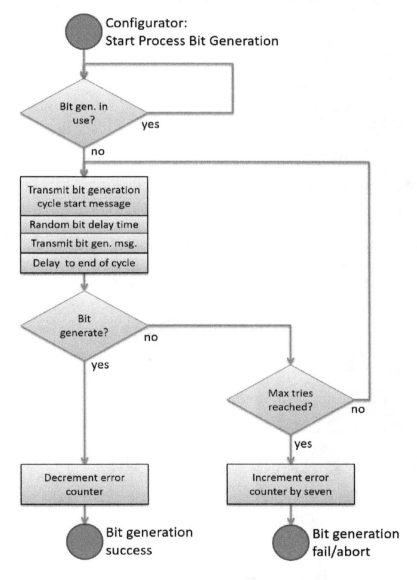

THE BIT-GENERATION CYCLE - CONFIGURATOR

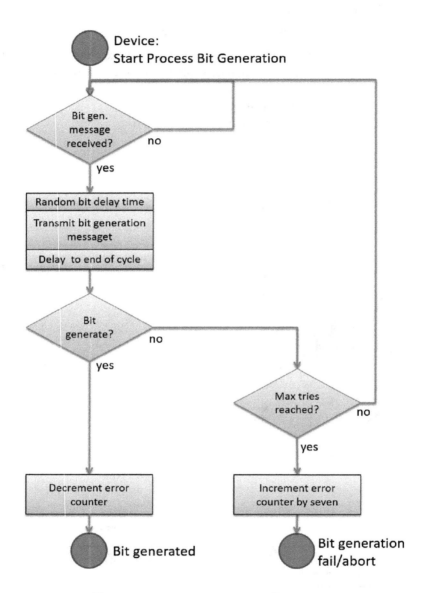

THE BIT-GENERATION CYCLE - DEVICE

7.2   CANcrypt system protocols

The protocols in this section are the backbone of CANcrypt. They include identification, generation of keys, and opening and closing procedures for secure channels.

7.2.1   Identification

The identification message verifies if the CANcrypt devices on the CAN system are compatible with each other.

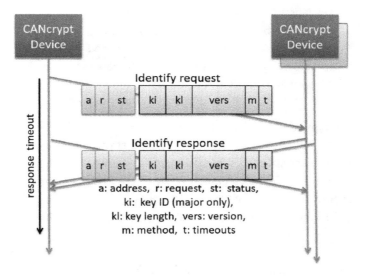

THE IDENTIFICATION REQUEST PROTOCOL

Parameters used for request (8-bit unless noted otherwise):

address (4 bits):	0, broadcast
request (4 bits):	3, identify
status:	current status byte
key ID:	key ID of the key that the manager requests to use
key length:	length of key
version (16 bits):	CANcrypt version number
method(4 bits):	method requested or supported
timeout(4 bits):	timeout requested or supported

The Identify request may come from any CANcrypt device and includes information about the key that should now be used and the version, method, and timeout. The channel number is set to the connection number of the device addressed (1–15). If set to zero, the identify request is a broadcast to all devices, and all devices must reply.

Parameters used for response (8-bit unless noted otherwise):

address (4 bits):   0, broadcast
request (4 bits):   identify
key ID: confirms the requested key ID
                         if not available, alternate key
key length: length of key
version (16 bits): CANcrypt version number
method (4 bits): confirms requested method
                         if not supported, alternate method
timeout (4 bits): confirms requested timeout
                         if not supported, alternate timeout

The CANcrypt device sends its response setting address to zero and copies its own version number into it. The key information, method, and timeout data is copied only if the device supports these parameters. A device that does not support a feature must modify the parameter in the response to indicate what the device can support. The requestor then has the option to send another identify request with different parameters, for example requesting to use a different key.

7.2.2 System startup and grouping

The CANcrypt system supports self-start for grouping with secure heartbeats. No specific manager or configurator is required to start the system. The devices repeatedly send startup requests that include a random number and how many partner devices they still see as missing. On reporting zero partners missing, all devices calculate the initial value for the dynamic key based on the last saved session key and the random numbers exchanged.

Each device then generates a startup acknowledgement message, which includes an authentication signature. On detecting a signature authentication failure, a device generates an alert. Otherwise all nodes start producing their secure heartbeats.

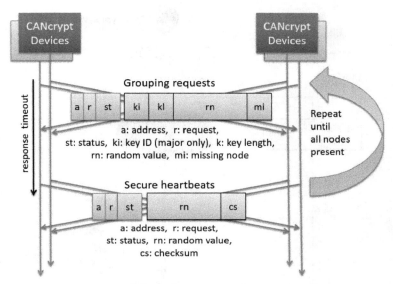

THE CANCRYPT GROUPING PROTOCOL

Parameters used for request (8-bit unless noted otherwise):

address (4 bits):   pairing address or 0 for grouping
request (4 bits):   4, grouping request
status: current status byte: grouping
key ID: key ID of the key that the manager requests to use
key length: length of key
rand (24 bits):   random value used for key initialization,
                    this value shall not change during the startup cycle
nodes missing: number of CANcrypt devices still missing

Once grouping is successful, all participating devices start producing the secure heartbeat.

The following flow charts show the processes involved. If a grouping request fails, then with each re-try an incremental delay is added. This delay shall increase exponentially to discourage hacking attempts that simply issue random re-tries over and over again.

If grouping fails due to different keys being requested, then the devices shall re-try using the highest priority key requested by any of the devices.

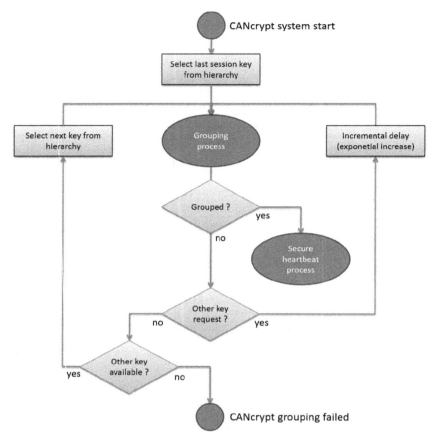

CANCRYPT GROUPING AND SECURE HEARTBEATS

Alternate CANopen use

For minimal CANopen integration, the grouping request can be embedded into the five manufacturer-specific bytes of the CANopen emergency message. The embedded data includes the status, the key ID and length, and the random number. The table below shows the recommended emergency codes to use.

EMCY code	Description
0091h	No error, grouping request, still nodes missing
0092h	No error, grouping request, no more nodes missing

The following flowchart shows the processes that execute during the grouping cycle in the CANcrypt devices.

Once grouping is complete, all grouped devices start producing their secure heartbeats.

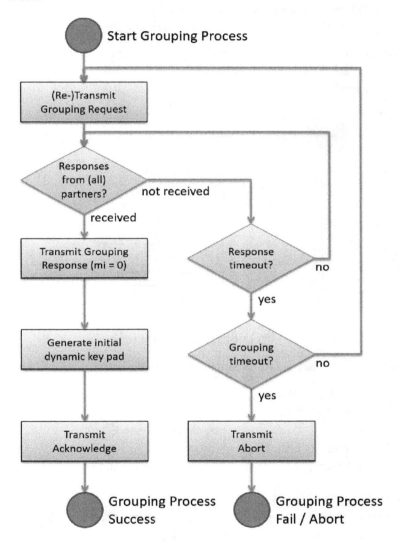

THE CANCRYPT GROUPING PROCESS

7.2.3 Secure heartbeat and dynamic key update

When the grouping process is complete, all grouped devices start the secure heartbeat protocol. All devices transmit their heartbeats in response to an initial heartbeat which is triggered by the device where the heartbeat event timer first expires. This mechanism ensures that all heartbeats are synchronized. The random value and the checksum are encrypted based on the current dynamic key.

All devices receive all heartbeats. At the end of each cycle (response timeout), the devices update their dynamic key by manipulating it based on the random values from all heartbeats.

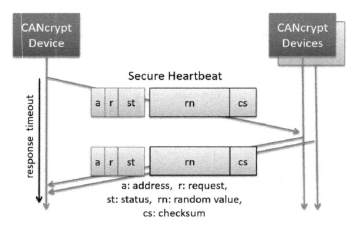

THE SECURE HEARTBEAT PROTOCOL

Parameters used for request (8-bit unless noted otherwise):

address (4 bits):	0, broadcast
request (4 bits):	9, secure heartbeat
status:	current status byte
random (24 bits):	random value
checksum:	checksum of status and random

Alternate CANopen use

For minimal CANopen integration, the secure heartbeat can be embedded into the 5 manufacturer specific bytes of the CANopen emergency message. The embedded data includes the random number and the checksum. The table below shows the recommended emergency codes to use.

EMCY code	Description
0093h	No error, CANcrypt secure heartbeat, all previous messages authenticated
8320h	Generic ungroup request, communication no longer secure
8330h	Alert: intruder detected by this device

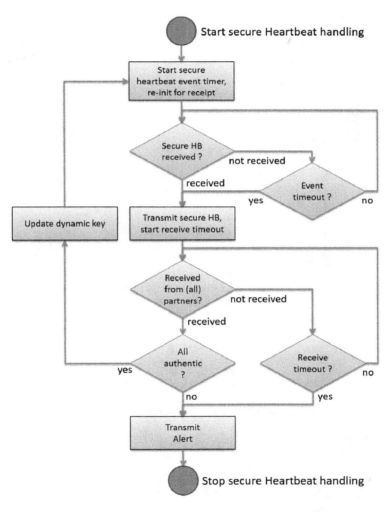

THE SECURE HEARTBEAT HANDLING PROCESS

The flowchart above illustrates how CANcrypt devices handle the secure heart-beat generation and verification. The process ensures that all devices synchronize themselves with each other. After one device produces the secure heartbeat, all others will produce heartbeats in response.

7.3   Secure messaging

Once devices are paired, secure message processing is enabled.

7.3.1   Secure message with preamble

On detecting a CAN message that requires protection (is in secure message list), the CANcrypt transmit handler generates a preamble that transmits before the data message.

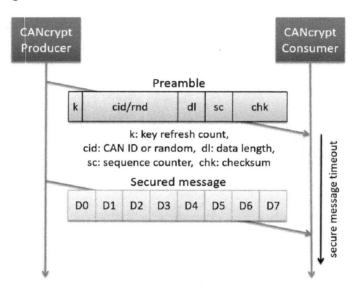

THE SECURE MESSAGE PROTOCOL

Parameters used for the preamble (8-bit unless noted otherwise):

key refresh (2 bits):	counter for key refresh cycles
ID or random (30 bits):	depending on use, repeat CAN ID or fill with random data
data length:	number of bytes in data message
seq. counter:	message counter
checksum (16 bits):	checksum covering preamble and data

The key refresh counter is incremented by all CANcrypt devices each time the shared dynamic key is updated. It therefore also synchronized between all devices. If the key counter is not encrypted, a receiver can verify if the message received was secured by the same active dynamic key that the receiver currently uses. If it does not match, then the receiver knows that it has to apply the dynamic key from the previous cycle as that message was still transmitted when that was active.

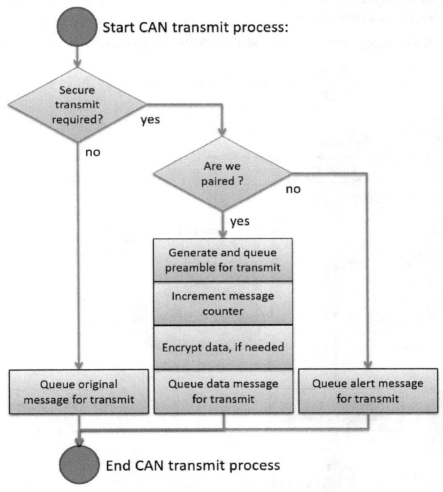

THE CAN MESSAGE TRANSMIT PROCESS

The configuration of the message list determines whether the following data message has encrypted bytes.

If a message to be transmitted is in the list of secure messages, the device first needs to verify that it is currently paired or grouped. If not, the device transmits an alert instead of the secure message.

A device that is paired or grouped (but NOT both) generates the preamble and depending on the configuration, the device may encrypt the preamble and message before adding them to the transmit queue.

The receiving process is illustrated by the following two flow charts.

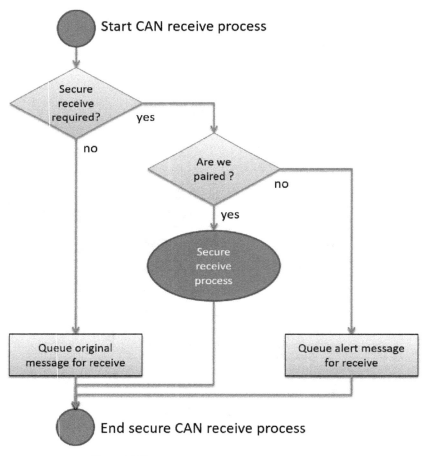

THE CAN MESSAGE RECEIVE PROCESS

If a message received is in the list of secure messages, the device first needs to verify that it is currently paired. If not, the device transmits an alert message.

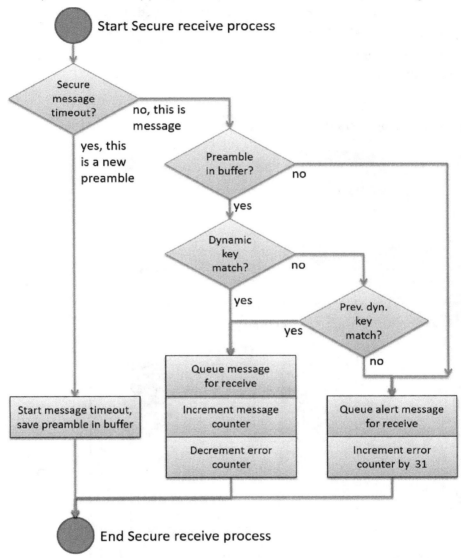

THE SECURE CAN MESSAGE RECEIVE PROCESS

A device that is paired starts the subprocess for secure reception. The device first checks if the secure message timeout has occurred. If so, the received data is the preamble, which the device must buffer until receiving the data message.

After receiving both the preamble and data message, the device applies the appropriate decryption and authentication algorithms using the current dynamic key. If authentication fails, the device retries the decryption and authentication using the previous dynamic key.

Retrying with the previous dynamic key ensures that a device can receive a message even if the transmitter has used the previous dynamic key.

7.3.2 Secure stream with preamble and epilog

CANcrypt supports secure message bursts or blocks of up to eight messages.

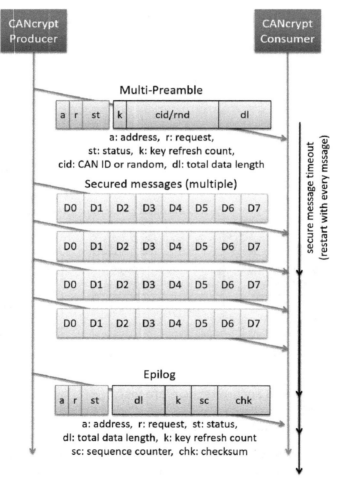

THE SECURE MESSAGE STREAM PROTOCOL

Parameters used for the preamble (8-bit unless noted otherwise):

address (4 bits):	0, broadcast
request (4 bits):	14, secure message preamble
key refresh (2 bits):	counter for key refresh cycles
ID or random (30 bits):	depending on use, repeat CAN ID or fill with random data
data length (16 bits):	total length of data or FFFFh for unknown

The configuration of the message list determines whether the following data messages have encrypted bytes.

Parameters used for epilog (8-bit unless noted otherwise):

address (4 bits):	0, broadcast
request (4 bits):	14, secure message preamble
data length (16 bits):	total number of bytes sent
key refresh:	counter for key refresh cycles
seq. counter:	message counter (incremented by number of message since preamble)
checksum (16 bits):	checksum covering preamble, data and epilog

7.3.3 Secure generic data object access

Secure generic data object access is a transfer mode that allows paired CANcrypt devices to directly exchange data using CANcrypt messages. A 16-bit index and 8-bit sub-index address the data within the devices. The mode is compatible with CANopen and can also be used generically to address data available in the devices.

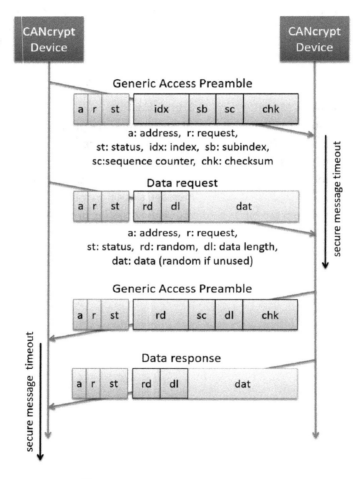

THE GENERIC ACCESS PROTOCOL

This transfer mode is also used by the configurator to erase, set, or save permanent keys. See section 7.5.3 for details.

Parameters used for generic access preamble request and response (8-bit unless noted otherwise):

address (4 bits):	1–15, address of paired partner
request (4 bits):	10, generic read access
	11, generic write access
status:	current status byte
index (16 bits):	index addressing the data in device
sub-index:	sub-index addressing the data in the device
seq. counter:	device's secure transmit counter
check (16 bits):	checksum of preamble and data

Parameters used for data request (8-bit unless noted otherwise):

address (4 bits):	1–15, address of paired partner
request (4 bits):	12, generic access data
random:	reserved, random value
data length:	length of data in bytes
data (32 bits):	on read: fill with random data,
	on write: data to write, fill unused bytes
	with random data

Parameters used for data response (8-bit unless noted otherwise):

address (4 bits):	1–15, address of paired partner
request (4 bits):	12, generic access data
random:	reserved, random value
data length:	length of data in bytes
data (32 bits):	on read: data requested, fill unused bytes
	with random data
	on write: fill with random data

7.4   Pairing

In addition to the grouping method, CANcrypt supports the pairing of two devices. Paired devices have an individual security channel. Pairing is intended for communication between the configurator and a device.

7.4.1   Pairing with a single device, open a channel

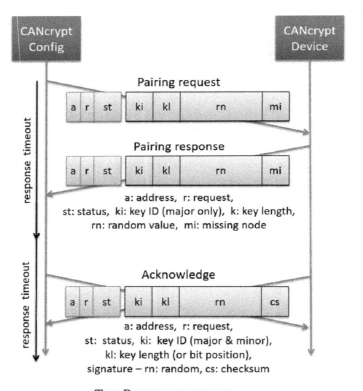

THE PAIRING PROTOCOL

The pairing protocol establishes a secure point-to-point communication based on the selected key. The process includes generating a pair dynamic key from the selected key.

Parameters used for request (8-bit unless noted otherwise):

address (4 bits):   1–15, device address to pair with
request (4 bits):   4, pairing
status: current status byte, pairing in process
key ID: key ID of key to use as a start
key length: length of this key
rand (24 bits):   random value used for key initialization,
                    this value shall not change during the startup cycle
node missing: 1

Parameters used for response (8-bit unless noted otherwise):

address (4 bits):   1–15, address of requesting device
request (4 bits):   4, pairing
status: current status byte, pairing in process
key ID: confirming key ID
key length: confirming length
rand (24 bits):   random value used for key initialization,
                    this value shall not change during the startup cycle
node missing: 0

The initiator starts with the pairing request message. Unless the key selected is the last session key, the address value must be in the range 1–15 to select one specific pairing partner.

The CANcrypt device addressed confirms the request by returning the request if the device has the key (ID and length) available. Otherwise the device replies with the next higher key ID and length available in the device, the protocol aborts, and the manager needs to start over).

The initiator now starts the subprotocol random-bit-generation to generate as many bits as needed to address any bit in the specified key. If the key length is 256 bits, eight bits are needed. If the key is 1024 bits, 10 bits are needed.

The dynamic key is initialized by copying bits from the selected key and applying a mixup function using the random value generated. This method ensures that the dynamic key is not always initialized with the same value.

7.4.2 Unpairing or closing a secure channel

Any paired device may request to close the secure connection at any time.

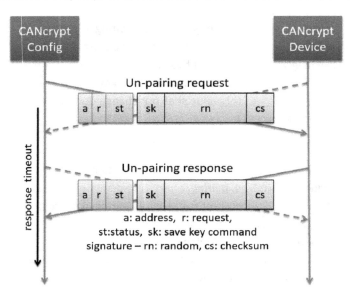

THE CLOSE SECURE CHANNEL PROTOCOL

Parameters used for request and response (8-bit unless noted otherwise):

address (4 bits):	1–15, device address
request (4 bits):	5, unpair
status:	current status byte
sign. (32 bit)	signature, 3 random bytes, 1 byte checksum, all encrypted based on dynamic key

7.5 Configurator protocols

The protocols described in this section are reserved for the CANcrypt configurator. They are primarily required for the key management involving generation of keys and erasing or saving them.

7.5.1 Extended Identification

The CANcrypt configurator has the option to request more detailed information using the identify request. To address the data, the index and sub-index system of

CANopen is used. Devices not implementing CANopen shall at least implement the identity object as defined in the table below.

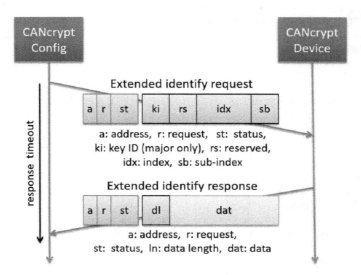

THE EXTENDED IDENTIFICATION PROTOCOL

Parameters used for extended request (8-bit unless noted otherwise):

address (4 bits):   1–15, device address
request (4 bits):   13, identify
status: current status byte
key ID: key ID requested
reserved: 0
index (16 bits):   index to information requested
sub-index: sub-index to information requested

Parameters used for extended request (8-bit unless noted otherwise):

address (4 bits):   1, to address the configurator
request (4 bits):   13, identify
status: current status byte
length: length of data in bytes (1–4)
data: data field

Mandatory identity information

Use	Type	Index	Sub-index
Vendor ID (0 if not CANopen Vendor)	UNSIGNED32	1018h	1
Product code	UNSIGNED32	1018h	2
Revision number	UNSIGNED32	1018h	3
Serial number (0 if not used)	UNSIGNED32	1018h	4

Optional identity information

Use	Type	Index	Sub-index
Manufacturer device name	STRING4	1008h	1–32
Manufacturer hardware version	STRING4	1009h	1–32
Manufacturer software version	STRING4	100Ah	1–32
CANopen communication data	1–4 bytes	1000h–1FFFh	any
Manufacturer specific data	1–4 bytes	2000h–5FFFh	any

7.5.2 Key generation or exchange

The CANcrypt configurator controls when new keys are generated and if they are saved. The maximum key size supported by this protocol is 32 bits. If larger keys need to be generated, multiple runs of this protocol are required.

The timeouts and methods used are those confirmed during the last identification protocol cycle.

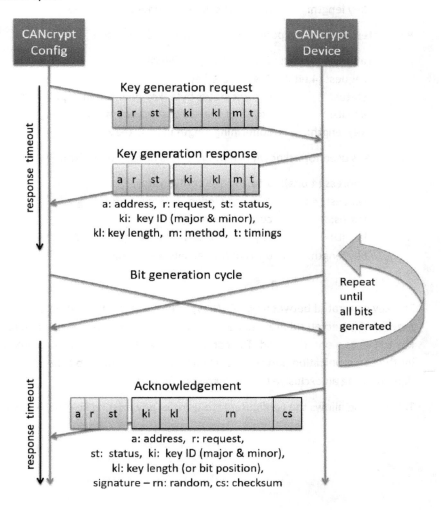

THE KEY GENERATION PROTOCOL

Parameters used for request (8-bit unless noted otherwise):

address (4 bits): 2–15, device address
request (4 bits): 7, generate key
                   8, trigger for bit-generation cycle
status: current status byte, key generation in progress
key ID: key ID of key that is generated
key length: length of this key segment in bits, 1–32

Parameters used for response (8-bit unless noted otherwise):

address (4 bits): 1, addressing configurator
request (4 bits): 7, generate key
status: current status byte, key generation in progress
key ID: key ID of the key that is generated
key length: confirming length in bits, 1–32

Parameters used for acknowledge (8-bit unless noted otherwise):

address (4 bits): 1 to address configurator
request (4 bits): 1, acknowledge
status: current status byte
key ID: key ID of the key that was generated
key length: 0 to confirm all bits were generated

7.5.3 Secret key transfer

The key generated between two CANcrypt devices is random. At times, devices may need to secretly share and transfer keys. To do so, a device can generate a key to use as a one-time pad. The configurator can use the flip-bits command to let the communication partner know which of the generated bits need to be flipped (using an exclusive OR) to get a specific key.

This method allows the configurator to transfer any existing key to the devices.

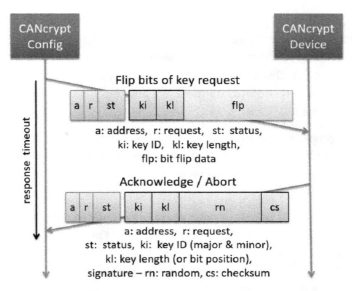

THE FLIP-KEY-BITS PROTOCOL

Parameters used for request (8-bit unless noted otherwise):

address (4 bits): 2–15: device address
request (4 bits): 6: flip bits
status: current status byte
key ID: key ID of key that is generated
key length: length of this key segment in bits, 1–32
flip data: an exclusive OR is executed with the last
                   generated key and this data

Parameters used for acknowledge (8-bit unless noted otherwise):

address (4 bits): 1 to address configurator
request (4 bits): 0: abort
                   1: acknowledge
status: current status byte
key ID: key ID of the key that needs bits flipped
key length: length of this key segment in bits, 1–32
sign. (32 bits)   signature, 3 random bytes, 1 byte checksum,
                   all encrypted based on dynamic key

7.5.4 Key access: erase, set and save

Only the configurator has access to the key management functions of erase, set, and save. The configurator must be actively paired with the device whose keys will be configured. Because saving and erasing keys affects security, these operations must use authenticated transfers, the secure generic data object access as described in section 7.3.3 .

The configurator can save a key only if the key is "erased" (all bits set) AND a key was newly generated (the key generation protocol was executed prior to the command). The key erase command works only on the current key (used to open the secure channel).

CANcrypt provides one entry (addressed by index and subindex) to select a key by its major and minor key IDs (see section 5.3.2 "The Keys").

Name	Index	Subindex	Type	Access	M/O
User key access	5EF1h	1	UNSIGNED8	R/W	O
Identifier for above	5EF2h	1	UNSIGNED32	R/W	O
Owner key access	5EF1h	2	UNSIGNED8	R/W	O
Identifier for above	5EF2h	2	UNSIGNED32	R/W	O
System integrator key access	5EF1h	3	UNSIGNED8	R/W	O
Identifier for above	5EF2h	3	UNSIGNED32	R/W	O
Manufacturer key access	5EF1h	4	UNSIGNED8	R/W	O
Identifier for above	5EF2h	4	UNSIGNED32	R/W	O

Reading the access field at 5EF1h provides the following information:

> 11h: no key stored at this location
>
> 33h: a key is stored at this location (see 5EF2h for identifier)
>
> FFh: this entry was never used before

On writing the access fields, the following values are accepted:

> DDh: delete the key stored at this location
>
> 5Dh: save the current dynamic key at this location

A CANcrypt device may deny write accesses depending on the key addressed and the configuration.

7.6   Protocol recordings

The following tables show CAN trace recordings of different key generation scenarios. The traces show the creation of a random 32-bit key using the different methods supported by CANcrypt. Debug mode is enabled so at the end of the cycle, both the configurator and the device show their secretly generated key.

Key Trace 1, below, uses a 100-ms cycle-time for each bit and the delay method for the bit-generation messages with a random delay of up to 64 ms. The first cycle starts at line 3, where the configurator generates the cycle start trigger message. Then both devices send their randomly selected bit-claim message. In lines 4 and 5, the messages are different (6F0 and 6F1), so the configurator generates a bit. As observers, we do not know which device transmitted which message so we do not which bit was generated.

The cycle beginning at line six illustrates that not every bit-cycle is successful. In lines 7 and 8, both devices chose the same bit-claiming message of 6F0h so the cycle repeats until the devices use different bit-claiming messages (lines 16 and 17).

Count	Time (ms)	ID	Message Type	Data (Hex)		Content
1	0.000	0x171	Cc Configurator	27 80 00 20		start 32bit key device 2
2	0.687	0x172	Cc Device 2	17 80 00 20		start 32bit key confirmation
3	101.971	0x171	Cc Configurator	07 80 20		start bit cycle - 32
4	154.733	0x6F0	Bit claim		52.762	delay
5	160.973	0x6F1	Bit claim		59.002	delay
6	202.967	0x171	Cc Configurator	07 80 1F		start bit cycle - 31
7	209.727	0x6F0	Bit claim		6.760	delay
8	257.984	0x6F0	Bit claim		55.017	delay
9	303.969	0x171	Cc Configurator	07 80 1F		start bit cycle - 31
10	334.079	0x6F1	Bit claim		30.110	delay
11	354.831	0x6F1	Bit claim		50.862	delay
12	405.081	0x171	Cc Configurator	07 80 1F		start bit cycle - 31
13	410.089	0x6F0	Bit claim		5.008	delay
14	437.841	0x6F0	Bit claim		32.760	delay
15	506.085	0x171	Cc Configurator	07 80 1F		start bit cycle - 31
16	545.085	0x6F1	Bit claim		39.000	delay
17	546.845	0x6F0	Bit claim		40.760	delay
18	607.078	0x171	Cc Configurator	07 80 1E		start bit cycle - 30
19	644.087	0x6F1	Bit claim		37.009	delay
20	656.848	0x6F0	Bit claim		49.770	delay
21	708.082	0x171	Cc Configurator	07 80 1D		start bit cycle - 29
22	731.850	0x6F0	Bit claim		23.768	delay
23	754.083	0x6F1	Bit claim		46.001	delay
24	809.085	0x171	Cc Configurator	07 80 1C		start bit cycle - 28
25	835.093	0x6F0	Bit claim		26.008	delay
26	871.847	0x6F1	Bit claim		62.762	delay
27	910.104	0x171	Cc Configurator	07 80 1B		start bit cycle - 27
28	929.856	0x6F1	Bit claim		19.752	delay
29	932.104	0x6F0	Bit claim		22.000	delay
30	1011.098	0x171	Cc Configurator	07 80 1A		start bit cycle - 26
31	1021.858	0x6F0	Bit claim		10.760	delay
32	1055.100	0x6F0	Bit claim		44.002	delay
33	1112.094	0x171	Cc Configurator	07 80 1A		start bit cycle - 26
34	1150.102	0x6F1	Bit claim		38.008	delay
35	1170.863	0x6F1	Bit claim		58.769	delay
...	...	...	...	...		...
186	6263.795	0x171	Cc Configurator	07 80 03		start bit cycle - 3
187	6269.555	0x6F1	Bit claim		5.760	delay
188	6279.747	0x6F0	Bit claim		15.952	delay
189	6364.898	0x171	Cc Configurator	07 80 02		start bit cycle - 2
190	6399.667	0x6F0	Bit claim		34.769	delay
191	6419.852	0x6F0	Bit claim		54.954	delay
192	6465.902	0x171	Cc Configurator	07 80 02		start bit cycle - 2
193	6490.678	0x6F0	Bit claim		24.776	delay
194	6517.847	0x6F1	Bit claim		51.945	delay
195	6566.913	0x171	Cc Configurator	07 80 01		start bit cycle - 1
196	6574.673	0x6F0	Bit claim		7.760	delay
197	6620.851	0x6F1	Bit claim		53.938	delay
198	6664.139	0x172	Cc Device 2	11 90 00 00		key generation ACK
199	6665.379	0x6E2	Debug Device 2	11 90 00 20 4B D9 58 F2		Debug info with key
200	6669.211	0x6E1	Debug Config	11 90 00 20 4B D9 58 F2		Debug info with key

KEY TRACE 1: 100MS CYCLE, 64MS DELAY MODE, 2 CLAIM MESSAGES

In lines 199 and 200, the generated key is in the last four bytes of the debug messages. Both devices generated the same value.

The code producing debug messages should never be present in final product releases. Finding a way to enable such messages in a released product would make a fine back door for any hacker...

Note that it took almost 200 CAN messages and almost seven seconds to generate the 32-bit key.

In Key Trace 2, below, the cycle time was reduced to 10 ms. Instead of a random delay, the bit-claim messages are transmitted immediately after the trigger message. In lines 7 and 21, both devices chose the same random bit-claim message. The bus and trace recording show the message only once because both devices transmitted the message at the same time and the messages overlay on the bus.

Compared to Key Trace 1, the number of cycles did not change drastically, but the much shorter cycle time means that it took less than 700 ms to generate the random shared 32-bit key.

Count	Time (ms)	ID	Message Type	Data (Hex)	Content
1	0.000	0x171	Cc Configurator	27 80 00 20	start 32bit key device 2
2	0.688	0x172	Cc Device 2	17 80 00 20	start 32bit key confirmation
3	11.960	0x171	Cc Configurator	07 80 20	start bit cycle - 32
4	12.351	0x6F0	Bit claim		
5	12.734	0x6F1	Bit claim		
6	22.967	0x171	Cc Configurator	07 80 1F	start bit cycle - 31
7	23.359	0x6F0	Bit claim		
8	33.967	0x171	Cc Configurator	07 80 1F	start bit cycle - 31
9	34.359	0x6F0	Bit claim		
10	34.743	0x6F1	Bit claim		
11	44.959	0x171	Cc Configurator	07 80 1E	start bit cycle - 30
12	45.352	0x6F0	Bit claim		
13	45.735	0x6F1	Bit claim		
14	55.960	0x171	Cc Configurator	07 80 1D	start bit cycle - 29
15	56.351	0x6F0	Bit claim		
16	56.734	0x6F1	Bit claim		
17	66.959	0x171	Cc Configurator	07 80 1C	start bit cycle - 28
18	67.351	0x6F0	Bit claim		
19	67.735	0x6F1	Bit claim		
20	77.975	0x171	Cc Configurator	07 80 1B	start bit cycle - 27
21	78.367	0x6F0	Bit claim		
22	88.967	0x171	Cc Configurator	07 80 1B	start bit cycle - 27
23	89.360	0x6F0	Bit claim		
24	89.743	0x6F1	Bit claim		
...	...	...	...	...	...
145	616.977	0x171	Cc Configurator	07 80 03	start bit cycle - 3
146	617.359	0x6F1	Bit claim		
147	627.975	0x171	Cc Configurator	07 80 03	start bit cycle - 3
148	628.359	0x6F1	Bit claim		
149	638.975	0x171	Cc Configurator	07 80 03	start bit cycle - 3
150	639.367	0x6F0	Bit claim		
151	649.975	0x171	Cc Configurator	07 80 03	start bit cycle - 3
152	650.368	0x6F0	Bit claim		
153	650.752	0x6F1	Bit claim		
154	660.976	0x171	Cc Configurator	07 80 02	start bit cycle - 2
155	661.367	0x6F0	Bit claim		
156	661.750	0x6F1	Bit claim		
157	671.990	0x171	Cc Configurator	07 80 01	start bit cycle - 1
158	672.381	0x6F0	Bit claim		
159	672.764	0x6F1	Bit claim		
160	682.324	0x172	Cc Device 2	11 90 00 00	key generation ACK
161	683.588	0x6E2	Debug Device 2	11 90 00 20 04 8C 7F 98	Debug info with key
162	684.525	0x6E1	Debug Config	11 90 00 20 04 8C 7F 98	Debug info with key

KEY TRACE 2: 10MS CYCLE, DIRECT RESPONSE MODE, 2 CLAIM MESSAGES

Key Trace 3, below, uses 16 bit-claiming messages instead of two. A bit-generation cycle fails only if both devices choose the same bit-claiming message.

This method significantly reduces the probability of a bit-claiming cycle to fail (from 0.5 to 0.0625) so fewer cycles must repeat.

Count	Time (ms)	ID	Message Type	Data (Hex)	Content
1	0.688	0x171	Cc Configurator	27 80 00 20	start 32bit key device 2
2	1.376	0x172	Cc Device 2	17 80 00 20	start 32bit key confirmation
3	12.649	0x171	Cc Configurator	07 80 20	start bit cycle - 32
4	13.041	0x6F9	Bit claim		
5	23.664	0x171	Cc Configurator	07 80 20	start bit cycle - 32
6	24.057	0x6F3	Bit claim		
7	24.449	0x6F9	Bit claim		
8	34.649	0x171	Cc Configurator	07 80 1F	start bit cycle - 31
9	35.040	0x6F5	Bit claim		
10	35.432	0x6FC	Bit claim		
11	45.640	0x171	Cc Configurator	07 80 1E	start bit cycle - 30
12	46.031	0x6F5	Bit claim		
13	46.431	0x6FF	Bit claim		
14	56.647	0x171	Cc Configurator	07 80 1D	start bit cycle - 29
15	57.039	0x6F7	Bit claim		
16	57.439	0x6FD	Bit claim		
17	67.647	0x171	Cc Configurator	07 80 1C	start bit cycle - 28
18	68.039	0x6F0	Bit claim		
19	68.431	0x6FE	Bit claim		
20	78.663	0x171	Cc Configurator	07 80 1B	start bit cycle - 27
21	79.046	0x6F2	Bit claim		
22	89.646	0x171	Cc Configurator	07 80 1B	start bit cycle - 27
23	90.029	0x6F1	Bit claim		
24	90.420	0x6FE	Bit claim		
...	...	...	...	...	...
96	364.656	0x171	Cc Configurator	07 80 03	start bit cycle - 3
97	365.047	0x6FB	Bit claim		
98	365.438	0x6FE	Bit claim		
99	375.654	0x171	Cc Configurator	07 80 02	start bit cycle - 2
100	376.037	0x6F2	Bit claim		
101	376.437	0x6FD	Bit claim		
102	386.662	0x171	Cc Configurator	07 80 01	start bit cycle - 1
103	387.054	0x6F3	Bit claim		
104	387.445	0x6F5	Bit claim		
105	396.510	0x172	Cc Device 2	11 90 00 00	key generation ACK
106	397.766	0x6E2	Debug Device 2	11 90 00 20 03 A8 6E DA	Debug info with key
107	398.957	0x6E1	Debug Config	11 90 00 20 03 A8 6E DA	Debug info with key

KEY TRACE 3: 10MS CYCLE, DIRECT RESPONSE MODE, 16 CLAIM MESSAGES

The trace shows just two repeated cycles (lines 4 and 21). It took less than 400 ms and about 100 CAN messages to generate the 32-bit random key.

Key Trace 4, below, shows debug messages only. The 32-bit key generation cycle is repeated over and over to measure how long it took each cycle to execute. The parameters used for the key generation are the same as in Key Trace 3. On average, it took less than 400 ms to generate the 32-bit key.

Assuming that this method would be used to generate or transfer a 256-bit key, we can expect that in most cases it would take less than 3.5 seconds.

Count	Time (ms)	ID	Message Type	Data (Hex)	Content
1		0x6E1	Debug Config	07 80 0A 00 00 00 00 00	Debug: start cycle
2	373	0x6E2	Debug Device 2	11 90 00 20 40 D8 F3 28	Debug info with key
3	1	0x6E1	Debug Config	11 90 00 20 40 D8 F3 28	Debug info with key
4		0x6E1	Debug Config	07 80 0A 00 00 00 00 00	Debug: start cycle
5	373	0x6E2	Debug Device 2	11 90 00 20 95 B4 BA 31	Debug info with key
6	1	0x6E1	Debug Config	11 90 00 20 95 B4 BA 31	Debug info with key
7		0x6E1	Debug Config	07 80 0A 00 00 00 00 00	Debug: start cycle
8	362	0x6E2	Debug Device 2	11 90 00 20 62 87 B1 EC	Debug info with key
9	1	0x6E1	Debug Config	11 90 00 20 62 87 B1 EC	Debug info with key
10		0x6E1	Debug Config	07 80 0A 00 00 00 00 00	Debug: start cycle
11	396	0x6E2	Debug Device 2	11 90 00 20 CB DA D0 C9	Debug info with key
12	1	0x6E1	Debug Config	11 90 00 20 CB DA D0 C9	Debug info with key
13		0x6E1	Debug Config	07 80 0A 00 00 00 00 00	Debug: start cycle
14	396	0x6E2	Debug Device 2	11 90 00 20 DF 5B D7 0A	Debug info with key
15	1	0x6E1	Debug Config	11 90 00 20 DF 5B D7 0A	Debug info with key
16		0x6E1	Debug Config	07 80 0A 00 00 00 00 00	Debug: start cycle
17	396	0x6E2	Debug Device 2	11 90 00 20 52 A9 88 CE	Debug info with key
18	1	0x6E1	Debug Config	11 90 00 20 52 A9 88 CE	Debug info with key
19		0x6E1	Debug Config	07 80 0A 00 00 00 00 00	Debug: start cycle
20	385	0x6E2	Debug Device 2	11 90 00 20 46 88 5A 3F	Debug info with key
21	1	0x6E1	Debug Config	11 90 00 20 46 88 5A 3F	Debug info with key
22		0x6E1	Debug Config	07 80 0A 00 00 00 00 00	Debug: start cycle
23	418	0x6E2	Debug Device 2	11 90 00 20 A9 4A 82 A3	Debug info with key
24	1	0x6E1	Debug Config	11 90 00 20 A9 4A 82 A3	Debug info with key
25		0x6E1	Debug Config	07 80 0A 00 00 00 00 00	Debug: start cycle
26	396	0x6E2	Debug Device 2	11 90 00 20 F9 33 85 1E	Debug info with key
27	1	0x6E1	Debug Config	11 90 00 20 F9 33 85 1E	Debug info with key
28		0x6E1	Debug Config	07 80 0A 00 00 00 00 00	Debug: start cycle
29	385	0x6E2	Debug Device 2	11 90 00 20 DD B0 50 88	Debug info with key
30	1	0x6E1	Debug Config	11 90 00 20 DD B0 50 88	Debug info with key
31		0x6E1	Debug Config	07 80 0A 00 00 00 00 00	Debug: start cycle
32	385	0x6E2	Debug Device 2	11 90 00 20 62 2E 2A C2	Debug info with key
33	1	0x6E1	Debug Config	11 90 00 20 62 2E 2A C2	Debug info with key
34		0x6E1	Debug Config	07 80 0A 00 00 00 00 00	Debug: start cycle
35	363	0x6E2	Debug Device 2	11 90 00 20 F1 6C 39 ED	Debug info with key
36	1	0x6E1	Debug Config	11 90 00 20 F1 6C 39 ED	Debug info with key
37		0x6E1	Debug Config	07 80 0A 00 00 00 00 00	Debug: start cycle
38	363	0x6E2	Debug Device 2	11 90 00 20 F0 D2 39 28	Debug info with key
39	1	0x6E1	Debug Config	11 90 00 20 F0 D2 39 28	Debug info with key
40		0x6E1	Debug Config	07 80 0A 00 00 00 00 00	Debug: start cycle
41	385	0x6E2	Debug Device 2	11 90 00 20 6D 67 46 9F	Debug info with key
42	1	0x6E1	Debug Config	11 90 00 20 6D 67 46 9F	Debug info with key

KEY TRACE 4: MULTIPLE RUN CYCLE TIMING

7.6.1 Grouping example

In a system where all CANcrypt devices support saving the last session key, the power-up boot cycle can be reduced to the following commands and responses:

Count	Time	ID	Message Type	Data (Hex)	Content
0	0	0x000	NMT Master Request		Reset
1	1	0x702	Bootup		
2	1	0x703	Bootup		
3	2	0x707	Bootup		
4	3	0x707	CANopen Heartbeat		Pre-Operational
5	4	0x703	CANopen Heartbeat		Pre-Operational
6	5	0x177	CC Device 7	04 04 02 84 0F A5 AF 03	Group request
7	6	0x173	Cc Device 3	04 04 02 84 FB 97 61 03	Group request
8	6	0x702	CANopen Heartbeat		Pre-Operational
9	7	0x172	Cc Device 2	04 04 02 84 CA 64 CB 03	Group request
10	33	0x172	Cc Device 2	04 08 02 84 CA 64 CB 00	Group confirmation
11	34	0x177	CC Device 7	04 08 02 84 0F A5 AF 00	Group confirmation
12	35	0x173	Cc Device 3	04 08 02 84 FB 97 61 00	Group confirmation
13	36	0x702	CANopen Heartbeat		Operational
14	36	0x703	CANopen Heartbeat		Operational
15	37	0x707	CANopen Heartbeat		Operational
16	284	0x177	CC Device 7	09 08 DF 7A 4A 02	Secure Heartbeat
17	285	0x172	Cc Device 2	09 08 FB E8 82 68	Secure Heartbeat
18	285	0x173	Cc Device 3	09 08 4B FB 3D E1	Secure Heartbeat
19	787	0x177	CC Device 7	09 08 C8 0F 87 48	Secure Heartbeat
20	788	0x172	Cc Device 2	09 08 ED 6F F0 93	Secure Heartbeat
21	789	0x173	Cc Device 3	09 08 FF 9D D3 89	Secure Heartbeat
22	1290	0x177	CC Device 7	09 08 9B A0 C5 C3	Secure Heartbeat
23	1291	0x173	Cc Device 3	09 08 B2 67 2B 36	Secure Heartbeat
24	1292	0x172	Cc Device 2	09 08 DE 98 25 7F	Secure Heartbeat
25	1794	0x177	CC Device 7	09 08 80 1F 9A 0A	Secure Heartbeat
26	1794	0x172	Cc Device 2	09 08 11 B5 32 EF	Secure Heartbeat
27	1795	0x173	Cc Device 3	09 08 1D F6 28 9D	Secure Heartbeat
28	2297	0x177	CC Device 7	09 08 83 F1 26 9B	Secure Heartbeat
29	2298	0x172	Cc Device 2	09 08 F2 BD 11 F6	Secure Heartbeat
30	2299	0x173	Cc Device 3	09 08 35 76 BA 5C	Secure Heartbeat
31	2801	0x177	CC Device 7	09 08 01 A6 15 F4	Secure Heartbeat
32	2802	0x173	Cc Device 3	09 08 DC 92 6C 2F	Secure Heartbeat
33	2802	0x172	Cc Device 2	09 08 C2 CE EC A1	Secure Heartbeat
34	3304	0x177	CC Device 7	09 08 87 47 90 E9	Secure Heartbeat
35	3305	0x172	Cc Device 2	09 08 5A 6B F4 DD	Secure Heartbeat
36	3306	0x173	Cc Device 3	09 08 4D 66 1C 4F	Secure Heartbeat

GROUPING TRACE: START-UP OF 3 GROUPED DEVICES

Here the devices 2, 3 and 7 all transmit their initial grouping request, followed by their grouping confirmation messages. Once grouping was established, all three devices start the secure heartbeat cycles.

7.6.2 Pairing example

Count	Time	ID	Message Type	Data (Hex)	Content
1	0	0x703	Bootup		
2	457	0x701	Bootup		
3	461	0x171	Cc Configurator	34 01 02 84 C1 FF 79 00	Initiate pairing request
4	462	0x173	Cc Device 3	04 01 02 84 63 C5 11 00	Pairing response
5	466	0x171	Cc Configurator	31 02 00 00 00 A2 64 19	Pairing confirmation
6	467	0x173	Cc Device 3	11 12 00 00 4B 49 45 D9	Pairing confirmation
7	476	0x171	Cc Configurator	37 82 00 01 91	Key update request: dynamic key
8	477	0x173	Cc Device 3	17 82 00 01 11	Key update response
9	503	0x171	Cc Configurator	08 82 01	Generate next bit
10	507	0x6FB	Bit claim		
11	517	0x6FA	Bit claim		
12	1531	0x171	Cc Configurator	08 02 01	Generate next bit
13	1532	0x6F9	Bit claim		
14	1537	0x6FA	Bit claim		
15	2002	0x183	Default: PDO	83 01 00 80 EC CB 9A BA	Preamble, TPDO 1, Node 3
16	2003	0x183	Default: PDO	A5 05 2D 2B 76 BC 6C 4F	Secured message, TPDO 1, Node 3
17	2005	0x283	Default: PDO	A5 05 01 07 5A	Decrypted echo data by configurator
18	2559	0x171	Cc Configurator	08 02 01	Generate next bit
19	2565	0x6FF	Bit claim		
20	2572	0x6F2	Bit claim		
21	2681	0x181	Default: PDO	81 01 00 C0 E1 C6 30 70	Preamble, TPDO 1, Configurator
22	2682	0x181	Default: PDO	33 03 29 2D E4 1F AE 2C	Secured message, TPDO 1, Configurator
23	2684	0x281	Default: PDO	33 03 01 05 CC	Decrypted echo data by node 3
24	2687	0x183	Default: PDO	83 01 00 C0 E1 C7 5D A9	Preamble, TPDO 1, Node 3
25	2688	0x183	Default: PDO	A5 05 2C 29 74 E4 55 4D	Secured message, TPDO 1, Node 3
26	2690	0x283	Default: PDO	A5 05 02 07 5A	Decrypted echo data by configurator
27	2779	0x183	Default: PDO	83 01 00 C0 E6 C7 91 F9	Preamble, TPDO 1, Node 4
28	2780	0x183	Default: PDO	A5 05 30 34 69 2F 08 BB	Secured message, TPDO 1, Node 4
29	2782	0x283	Default: PDO	A5 05 03 07 5A	Decrypted echo data by configurator
30	3126	0x181	Default: PDO	81 01 00 C0 E1 C7 BD B8	Preamble, TPDO 1, Configurator
31	3127	0x181	Default: PDO	33 03 2C 2B E2 D3 37 0D	Secured message, TPDO 1, Configurator
32	3129	0x281	Default: PDO	33 03 02 05 CC	Decrypted echo data by node 3
33	3132	0x183	Default: PDO	83 01 00 C0 F7 D7 ED E4	Preamble, TPDO 1, Node 4
34	3133	0x183	Default: PDO	A5 05 2C 2F 72 93 F2 26	Secured message, TPDO 1, Node 4
35	3135	0x283	Default: PDO	A5 05 04 07 5A	Decrypted echo data by configurator
36	3556	0x183	Default: PDO	83 01 00 C0 F4 D7 8A 2F	Preamble, TPDO 1, Node 5
37	3557	0x183	Default: PDO	A5 05 28 2A 77 AC 09 28	Secured message, TPDO 1, Node 5
38	3559	0x283	Default: PDO	A5 05 05 07 5A	Decrypted echo data by configurator
39	3571	0x181	Default: PDO	81 01 00 C0 E6 C7 0A 31	Preamble, TPDO 1, Configurator
40	3572	0x181	Default: PDO	33 03 30 36 FF 74 7F 5C	Secured message, TPDO 1, Configurator
41	3574	0x281	Default: PDO	33 03 03 05 CC	Decrypted echo data by node 3
42	3577	0x183	Default: PDO	83 01 00 C0 F6 D4 29 6E	Preamble, TPDO 1, Node 5
43	3577	0x183	Default: PDO	A5 05 29 28 75 0B A7 37	Secured message, TPDO 1, Node 5
44	3580	0x283	Default: PDO	A5 05 06 07 5A	Decrypted echo data by configurator
45	3587	0x171	Cc Configurator	08 02 01	Generate next bit
46	3589	0x6FC	Bit claim		
47	3589	0x6FE	Bit claim		
48	4016	0x181	Default: PDO	81 01 00 00 11 31 F6 EA	Preamble, TPDO 1, Configurator
49	4017	0x181	Default: PDO	33 03 60 61 A8 8A C8 A0	Secured message, TPDO 1, Configurator
50	4019	0x281	Default: PDO	33 03 04 05 CC	Decrypted echo data by node 3
51	4022	0x183	Default: PDO	83 01 00 00 6B 46 04 76	Preamble, TPDO 1, Node 6
52	4023	0x183	Default: PDO	A5 05 67 67 3A 4D 45 B4	Secured message, TPDO 1, Node 6
53	4025	0x283	Default: PDO	A5 05 07 07 5A	Decrypted echo data by configurator

PAIRING TRACE: PAIRING AND SECURE MESSAGES

The trace above shows a configurator pairing itself with device 3. After the initial pairing request and confirmation, the configurator starts additional bit-generation cycles to update he shared dynamic key.

At line 15, device 3 sends a first secure message, first the preamble, followed by the data message. The message in line 17 is a test-only echo of the decrypted clear text message.

What follows is a combination of secure messages transmitted from both devices, the decrypted test-only clear echo data and further bit-generation cycles to update the dynamic key.

7.6.3 Secure Grouping example

Grouping can also be combined with secure messaging, as shown by the trace below.

Count	ID	Message Type	Data (Hex)	Details
1	0x702	Bootup Device 2	00	
2	0x702	Group Init Device 2	7F	Pre-Operational
3	0x172	Cc Device 2 group	04 04 02 82 5C 69 32 03	request
4	0x703	Bootup Device 3	00	
5	0x703	Group Init Device 3	7F	Pre-Operational
6	0x173	Cc Device 3 group	04 04 02 82 18 11 01 03	request
7	0x707	Bootup Device 7	00	
8	0x707	Group Init Device 7	7F	Pre-Operational
9	0x177	CC Device 7 group	04 04 02 82 22 57 B8 03	request
10	0x172	Cc Device 2 group	04 08 02 82 5C 69 32 00	confirmation
11	0x702	Heartbeat 2	05	Operational
12	0x173	Cc Device 3 group	04 08 02 82 18 11 01 00	confirmation
13	0x177	CC Device 7 group	04 08 02 82 22 57 B8 00	confirmation
14	0x703	Heartbeat 3	05	Operational
15	0x707	Heartbeat 7	05	Operational
16	0x172	Cc Device 2 secure Heartbeat	09 08 8A 73 CB 94	
17	0x173	Cc Device 3 secure Heartbeat	09 08 38 29 C5 8E	
18	0x177	CC Device 7 secure Heartbeat	09 08 A1 FD F3 16	
19	0x702	Heartbeat confirmed	05	Operational
20	0x703	Heartbeat confirmed	05	Operational
21	0x707	Heartbeat confirmed	05	Operational
22	0x177	CC Device 7 secure Heartbeat	09 08 7C A0 3F 9E	
23	0x172	Cc Device 2 secure Heartbeat	09 08 C4 B3 EC CE	
24	0x173	Cc Device 3 secure Heartbeat	09 08 D4 E7 F0 ED	
25	0x702	Heartbeat confirmed	05	Operational
26	0x703	Heartbeat confirmed	05	Operational
27	0x707	Heartbeat confirmed	05	Operational

SECURE GROUPING TRACE: INITIALIZATION

The startup and initialization phase is the same as with regular grouping. The three devices 2, 3 and 7 boot up, group and start producing the secure heartbeat cycles.

Count	ID	Message Type	Data (Hex)	Details
28	0x482	Device 2 data	33 02 01 02 CC	Copy of new message to send
29	0x182	Default: PDO	82 01 00 00 97 D1 95 F9	CANcrypt secure message
30	0x182	Default: PDO	33 2C B1 D7 80 40 A7 4B	CANcrypt secure message
31	0x283	Device 3 data	33 02 01 02 CC	Decrypted confirmation echo
32	0x287	Device 7 data	33 02 01 02 CC	Decrypted confirmation echo
33	0x483	Device 3 data	33 03 01 03 CC	Copy of new message to send
34	0x183	Default: PDO	83 01 00 00 D1 40 37 C6	CANcrypt secure message
35	0x183	Default: PDO	33 15 7A 6B 76 B1 AC 9A	CANcrypt secure message
36	0x287	Device 7 data	33 03 01 03 CC	Decrypted confirmation echo
37	0x282	Device 2 data	33 03 01 03 CC	Decrypted confirmation echo
38	0x487	Device 7 data	33 07 01 07 CC	Copy of new message to send
39	0x187	Default: PDO	87 01 00 00 C6 93 18 60	CANcrypt secure message
40	0x187	Default: PDO	33 ED F0 BB EC 8E 4D 69	CANcrypt secure message
41	0x282	Device 2 data	33 07 01 07 CC	Decrypted confirmation echo
42	0x283	Device 3 data	33 07 01 07 CC	Decrypted confirmation echo
43	0x482	Device 2 data	33 02 02 02 CC	Copy of new message to send
44	0x182	Default: PDO	82 01 00 00 F0 5C F2 AA	CANcrypt secure message
45	0x182	Default: PDO	33 56 41 3E E0 C7 16 82	CANcrypt secure message
46	0x283	Device 3 data	33 02 02 02 CC	Decrypted confirmation echo
47	0x287	Device 7 data	33 02 02 02 CC	Decrypted confirmation echo
48	0x483	Device 3 data	33 03 02 03 CC	Copy of new message to send
49	0x183	Default: PDO	83 01 00 00 D9 24 46 5A	CANcrypt secure message
50	0x183	Default: PDO	33 DC 9D 07 79 C1 77 13	CANcrypt secure message
51	0x287	Device 7 data	33 03 02 03 CC	Decrypted confirmation echo
52	0x282	Device 2 data	33 03 02 03 CC	Decrypted confirmation echo
53	0x487	Device 7 data	33 07 02 07 CC	Copy of new message to send
54	0x187	Default: PDO	87 01 00 00 91 2C A2 EA	CANcrypt secure message
55	0x187	Default: PDO	33 39 F1 CA BA E7 0E 72	CANcrypt secure message
56	0x282	Device 2 data	33 07 02 07 CC	Decrypted confirmation echo
57	0x283	Device 3 data	33 07 02 07 CC	Decrypted confirmation echo
58	0x173	Cc Device 3 secure Heartbeat	09 08 C6 27 C1 B8	
59	0x177	CC Device 7 secure Heartbeat	09 08 4E 2B EE 40	
60	0x172	Cc Device 2 secure Heartbeat	09 08 57 A8 53 26	
61	0x702	Heartbeat confirmed		Operational
62	0x703	Heartbeat confirmed		Operational
63	0x707	Heartbeat confirmed		Operational

SECURE GROUPING TRACE: SECURE MESSAGING

In the example shown above, devices first transmit a copy of the secure data (so that we can see it) using CAN ID 48xh. Then they transmit the secure version at 18xh. First the preamble, then the encrypted data (in this configuration first byte is not encrypted).

8 CANcrypt Programming

The following diagram illustrates the simplified integration of CANcrypt into existing CAN systems. Integration happens at the driver level and thus is independent from additional layers and the software using CAN communications. Applications need only make a few function calls to activate the CANcrypt security system and to handle call backs from events reported by the CANcrypt security system.

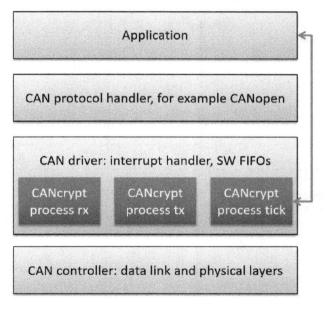

SIMPLIFIED CANCRYPT INTEGRATION

8.1   C include files definitions

The example demo projects provided contain the C include/definition files used for all major CANcrypt definitions and settings. See Appendix A for more details on the examples available and Appendix B for selected code listings.

8.1.1   CC_user_config.h

This module is used to configure the CANcrypt operation mode, including the selected methods and timeouts.

```
/********************************************************************
MODULE:     Cc_user_config.h, CANcrypt global user configuration
CONTAINS:   Configuration parameters for CANcrypt
AUTHOR:     2017 Embedded Systems Academy, GmbH
HOME:       www.esacademy.com/cancrypt

  Licensed under the Apache License, Version 2.0 (the "License");
  you may not use this file except in compliance with the License.
  You may obtain a copy of the License at
  www.apache.org/licenses/LICENSE-2.0

  Unless required by applicable law or agreed to in writing, software
  distributed under the License is distributed on an "AS IS" BASIS,
  WITHOUT WARRANTIES OR CONDITIONS OF ANY KIND, either express or
  implied. See the License for the specific language governing
  permissions and limitations under the License.

VERSION:    0.10, 19-JAN-2017
********************************************************************/

#ifndef _CC_USER_CONFIG_H
#define _CC_USER_CONFIG_H


/********************************************************************
CANcrypt code selection
********************************************************************/

// If defined, monitor CAN for unexpected messages
#define Cc_USE_MONITORING

// If defined, implement grouping functionality
#define Cc_USE_GROUPING

// If defined, implement key generation and pairing functionality
#define Cc_USE_PAIRING

// If defined, implement secure messaging
#define Cc_USE_SECURE_MSG

// If defined, switch output pins for performance measurements
//#define Cc_USE_DIGOUT
```

Method, key and timings

This section configures the security level used, the key length for the permanent and the dynamic key and the timings and timouts.

```
/******************************************************************
Security CAN Functionality
Cc_SECFCT_BASIC            0x00
Cc_SECFCT_REGULAR          0x01
Cc_SECFCT_ADVANCED         0x02
Cc_SECFCT_CUSTOM           0x03
******************************************************************/
#define Cc_FUNCTIONALITY        Cc_SECFCT_REGULAR


/******************************************************************
Permanent key length used by this version: 128, 256, 512 or 1024
Must be greater or equal to the dynamic key length
******************************************************************/
#define Cc_PERMKEY_LEN_BITS     128
#define Cc_PERMKEY_LEN32        (Cc_PERMKEY_LEN_BITS >> 5)
#define Cc_PERMKEY_LEN8         (Cc_PERMKEY_LEN_BITS >> 3)


/******************************************************************
Dynamic key length used by this version: 64 or 128
******************************************************************/
#define Cc_KEY_LEN_BITS         64
#define Cc_KEY_LEN32            (Cc_KEY_LEN_BITS >> 5)
#define Cc_KEY_LEN8             (Cc_KEY_LEN_BITS >> 3)

// Default length for key/bit-generation cycle
#define Cc_KEY_LEN_INIT         32


/******************************************************************
Timings used
IN THIS VERSION, INDIVIDUAL TIMINGS MUST STILL BE SET BELOW
******************************************************************/
#define Cc_TIMINGS              Cc_TIMING_MEDIUM


/******************************************************************
Secure heartbeat timings
******************************************************************/
// Secure Heartbeat event time
#define Cc_SECHB_EVENT_TIME     500

// Secure Heartbeat inhibit time
#define Cc_SECHB_INHIBIT_TIME   250

// Secure Heartbeat timeout
#define Cc_SECHB_TIMEOUT        1000

// Secure message timeout
#define Cc_SECMSG_TIMEOUT       20
```

Bit / key generation

The parameters in this section control the bit-generation cycles

```
/**********************************************************************
Key generation parameters
**********************************************************************/
// Bit-generation cycle timeout: 10, 25, 50, 100
#define Cc_BIT_CYCLE_TIMEOUT    25

// Bit generation max random delay time, 0 for immediate/direct
//                              or 0x0F, 0x1F, 0x3F
#define Cc_BIT_CYCLE_RANDTIME    0x0F

// Bit generation method used: random delay or direct
#define Cc_BITMETHOD            Cc_BITMETHOD_DELAY

// Number of bit claiming messages used: 2 (default) or 16
#define Cc_BITMETHOD_CLAIMS     Cc_BITMETHOD_16CLAIMS

// Maximum number of ignore states allowed for bit select cycles
#define Cc_MAX_ignore_cnt        8

// CANcrypt Method combination
#define Cc_METHOD               Cc_TIMINGS + \
    (((Cc_BITMETHOD + Cc_BITMETHOD_CLAIMS + Cc_FUNCTIONALITY) << 4))
```

Unexpected message monitoring

The mechanism used for monitoring the CANbus for unexpected messages uses a list of all CAN IDs requiring monitoring. Define the maximum length of this list here.

```
/**********************************************************************
Enable monitoring of unexpected CAN message IDs received.
**********************************************************************/
// Maximum number of CAN IDs in list monitored, 0 to disable
#define Cc_CANIDLIST_LEN        16
```

CAN message IDs used

Here we define the CAN message IDs used by CANcrypt.

```
/**********************************************************************
CAN IDs used
**********************************************************************/
// CANcrypt messages for devices and configurator, plus next 15 IDs
#define Cc_CANID_CONFIG         0x0171
// Bit claiming messages start with this ID, plus next 1 or 15 IDs

#define Cc_CANID_BITBASE        0x06F0

// CANcrypt messages for debug messages, plus next 15 IDs
```

```
#define Cc_CANID_DEBUG          0x06E1


/****************************************************************
DEFINES: CAN HARDWARE DRIVER DEFINITIONS
****************************************************************/

// Tx FIFO depth (must be 0, 4, 8, 16 or 32)
#define TXFIFOSIZE              16

// Rx FIFO depth (must be 0, 4, 8, 16 or 32)
#define RXFIFOSIZE              16

#endif
/*---------------------- END OF FILE ----------------------------*/
```

8.1.2 CANcrypt_types.h

This definition file contains the main bit and type definitions for the CANcrypt parameters, variables and configurations.

```
/****************************************************************
MODULE:     CANcrypt_types.h, global type definitions
CONTAINS:   The global type definitions required for CANcrypt
AUTHOR:     2017 Embedded Systems Academy, GmbH
HOME:       www.esacademy.com/cancrypt

   Licensed under the Apache License, Version 2.0 (the "License");
   you may not use this file except in compliance with the License.
   You may obtain a copy of the License at
   www.apache.org/licenses/LICENSE-2.0

   Unless required by applicable law or agreed to in writing, software
   distributed under the License is distributed on an "AS IS" BASIS,
   WITHOUT WARRANTIES OR CONDITIONS OF ANY KIND, either express or
   implied. See the License for the specific language governing
   permissions and limitations under the License.

VERSION:    0.10, 19-JAN-2017
****************************************************************/


#ifndef _CANCRYPT_TYPES_H
#define _CANCRYPT_TYPES_H

#include "Cc_user_types.h"
#include "Cc_user_config.h"
#include "CANcrypt_can.h"
```

CANcrypt version

```
/*********************************************************************
CANcrypt 16bit version info
6bit version, 6bit revision, 2bit reserved (zero), 2bit functionality
Version zero reserved for demo and evaluation
*********************************************************************/
#define Cc_VERSION_NR  0
#define Cc_REVISION_NR 1
#define CANcrypt_VERSION ( (Cc_VERSION_NR << 10) + \
                           (Cc_REVISION_NR << 4) + Cc_FUNCTIONALITY )
```

CANcrypt key hierarchy

```
/*********************************************************************
CANcrypt key hierarchy
Book section 5.3.2 "The Keys"
*********************************************************************/
#define Cc_PERM_KEY_MANUFACTURER   0x06
#define Cc_PERM_KEY_INTEGRATOR     0x05
#define Cc_PERM_KEY_OWNER          0x04
#define Cc_PERM_KEY_USER           0x03
#define Cc_PERM_KEY_SESSION        0x02
#define Cc_DYN_KEY_GROUP           0x01
#define Cc_DYN_KEY_PAIR            0x00
```

CANcrypt status byte

```
/*********************************************************************
CANcrypt status byte
Book section 5.3.3 "Status"
*********************************************************************/
// Bits 0-1: status of pairing process
#define Cc_PAIR_STAT_BITS          0x03
#define Cc_PAIR_STAT_NONE          0x00
#define Cc_PAIR_STAT_PROGRESS      0x01
#define Cc_PAIR_STAT_PAIRED        0x02
#define Cc_PAIR_STAT_FAIL          0x03
// Bits 2-3: status of grouping process
#define Cc_GROUP_STAT_BITS         0x0C
#define Cc_GROUP_STAT_NONE         0x00
#define Cc_GROUP_STAT_PROGRESS     0x04
#define Cc_GROUP_STAT_GROUPED      0x08
#define Cc_GROUP_STAT_FAIL         0x0C
// Bits 4-5: status of last command
#define Cc_CMD_STAT_BITS           0x30
#define Cc_CMD_STAT_NONE           0x00
#define Cc_CMD_STAT_SUCCESS        0x10
#define Cc_CMD_STAT_IGNORE         0x20
#define Cc_CMD_STAT_FAIL           0x30
// Bit 6: reserved
// Bit 7: status of key generation
#define Cc_KEY_STAT_BITS           0x80
#define Cc_KEY_STAT_NONE           0x00
#define Cc_KEY_STAT_GENERATION     0x80
```

CANcrypt controls

```
/*********************************************************************
CANcrypt request, response, event identification
Book section 5.3.4 "Controls"
*********************************************************************/
#define Cc_CTRL_BITS            0x0F
#define Cc_CTRL_ABORT           0x00
#define Cc_CTRL_ACK             0x01
#define Cc_CTRL_ALERT           0x02
#define Cc_CTRL_IDENTIFY        0x03
#define Cc_CTRL_PAIR            0x04
#define Cc_CTRL_UNPAIR          0x05
#define Cc_CTRL_FLIPBIT         0x06
#define Cc_CTRL_KEYGEN          0x07
#define Cc_CTRL_BITNOW          0x08
#define Cc_CTRL_SECHB           0x09
#define Cc_CTRL_DATRD           0x0A
#define Cc_CTRL_DARWR           0x0B
#define Cc_CTRL_DATMSG          0x0C
#define Cc_CTRL_XTDID           0x0D
#define Cc_CTRL_RESERVED        0x0E
#define Cc_CTRL_SAVEKEY         0x0F
```

CANcrypt methods

```
/*********************************************************************
CANcrypt security functionality
Book section 5.3.5 "Methods"
*********************************************************************/
// Security functionality: basic, regular, advanced or custom
#define Cc_SECFCT_BITS          0x03
#define Cc_SECFCT_BASIC         0x00
#define Cc_SECFCT_REGULAR       0x01
#define Cc_SECFCT_ADVANCED      0x02
#define Cc_SECFCT_CUSTOM        0x03
// Bit generation method: random delay or direct response
#define Cc_BITMETHOD_DIRECT     0x00
#define Cc_BITMETHOD_DELAY      0x04
// Bit claiming messages: 2 or 16
#define Cc_BITMETHOD_2CLAIMS    0x00
#define Cc_BITMETHOD_16CLAIMS   0x08
```

CANcrypt timings

```
/*********************************************************************
CANcrypt timings and timeouts
Book section 5.3.7 "Timings"
*********************************************************************/
#define Cc_TIMING_BITS          0x03
#define Cc_TIMING_FAST          0x00
#define Cc_TIMING_MEDIUM        0x01
#define Cc_TIMING_SLOW          0x02
#define Cc_TIMING_CUSTOM        0x03
```

CANcrypt events

```
/*********************************************************************
CANcrypt call back event and alert / error codes
Below 0x80: status, event  -  0x80 and higher: error, alert
*********************************************************************/
#define Cc_EVENT_GENSTAT          0x00
#define Cc_EVENT_KEYSTAT          0x10
#define Cc_EVENT_PARSTAT          0x20
#define Cc_EVENT_GRPSTAT          0x30
#define Cc_EVENT_MSGSTAT          0x40
#define Cc_EVENT_GASTAT           0x50
#define Cc_EVENT_GENERR           0x80
#define Cc_EVENT_KEYERR           0x90
#define Cc_EVENT_PARERR           0xA0
#define Cc_EVENT_GRPERR           0xB0
#define Cc_EVENT_MSGERR           0xC0
#define Cc_EVENT_GAERR            0xD0
```

CANcrypt alerts and errors

```
// new key generation initated
#define Cc_EVENT_KEYGEN_INIT      Cc_EVENT_KEYSTAT + 1
// key generation completed OK
#define Cc_EVENT_KEYGEN_OK        Cc_EVENT_KEYSTAT + 2

// new pairing started
// call-back parameters: key id, key length
#define Cc_EVENT_PAIR_INIT        Cc_EVENT_PARSTAT + 1
// key generation completed OK
#define Cc_EVENT_PAIRED           Cc_EVENT_PARSTAT + 2

// new grouping cycle
// call-back parameters: group info, HB event time, HB inhibit time
#define Cc_EVENT_GROUP_INIT       Cc_EVENT_GRPSTAT + 1
// grouping completed
// call-back parameters: group info, HB event time, HB inhibit time
#define Cc_EVENT_GROUPED          Cc_EVENT_GRPSTAT + 2
// secure heartbeat cycle complete
#define Cc_EVENT_HBSECURED        Cc_EVENT_GRPSTAT + 3

// generic read access requested (device only)
// call-back parameters: index+subindex
#define Cc_GACC_READ_REQUEST      Cc_EVENT_GASTAT + 1
// generic read access completed (configurator only)
// call-back parameters: device, data length, data
#define Cc_GACC_READ_OK           Cc_EVENT_GASTAT + 2
// generic write access requested (device only)
// call-back parameters: index+subindex, data length, data
#define Cc_GACC_WRITE_REQUEST     Cc_EVENT_GASTAT + 3
// generic write access completed (configurator only)
// call-back parameters: device, status
#define Cc_GACC_WRITE_OK          Cc_EVENT_GASTAT + 4
```

```
// unpairing event with instruction to
// save current dynamic key as session key
#define Cc_UNPAIR_SESSION       Cc_EVENT_GENSTAT + 1;
// save current dynamic key as session key
#define Cc_UNPAIR_USER          Cc_EVENT_GENSTAT + 2;

// intruder alert
#define Cc_ERR_INTRUDER         Cc_EVENT_GENERR + 1

// timeout in key generation
// call-backparameters: key work, key id, key position
#define Cc_ERR_KEY_GEN_TIME     Cc_EVENT_KEYERR + 1
// different bit expected in key generation
// call-backparameters: key work, key id, key position
#define Cc_ERR_KEY_GEN_BIT      Cc_EVENT_KEYERR + 2
// too many bits has to be ignored
// call-back parameters: key work, key id, key position
#define Cc_ERR_KEY_GEN_IGNORE   Cc_EVENT_KEYERR + 3

// trying to group and pair at the same time
#define Cc_ERR_PAIRGOUPED       Cc_EVENT_PARERR + 1

// grouping failed before secure heartbeat started
#define Cc_ERR_GROUP_SESS       Cc_EVENT_GRPERR + 1
// group partner lost
// call-back parameters: group expected, group found, 0
#define Cc_ERR_GROUP_LOST       Cc_EVENT_GRPERR + 2
// secure heartbeat failure
// call-back parameters: device with sec HB failure, 0, 0
#define Cc_ERR_SECHB_FAIL       Cc_EVENT_GRPERR + 3

// unexpected message failure
// call-back parameters: msg id
#define Cc_ERR_MSG_UNEXPECT     Cc_EVENT_MSGERR + 1
// secure message failure, not paired
// call-back parameters: msg id
#define Cc_ERR_MSG_NOPAIR       Cc_EVENT_MSGERR + 2
// secure message failure, not secure
// call-back parameters: msg id
#define Cc_ERR_MSG_NOSEC        Cc_EVENT_MSGERR + 3

// generic read access failed
#define Cc_GACC_READ_FAIL       Cc_EVENT_GAERR + 1
// generic read access timeout
#define Cc_GACC_READ_TIME       Cc_EVENT_GAERR + 2
// generic write access failed
#define Cc_GACC_WRITE_FAIL      Cc_EVENT_GAERR + 3
// generic write access timeout
#define Cc_GACC_WRITE_OK_TIME   Cc_EVENT_GAERR + 4
```

Restart parameters

```
/*********************************************************************
CANcrypt control for Cc_Restart
*********************************************************************/
// Bits 0-1: control for pairing process
#define Cc_PAIR_CTRL_NONE           0x00
#define Cc_PAIR_CTRL_RESTART        0x01
#define Cc_PAIR_CTRL_STOP           0x02
// Bits 2-3: control for grouping process
#define Cc_GROUP_CTRL_NONE          0x00
#define Cc_GROUP_CTRL_RESTART       0x04
#define Cc_GROUP_CTRL_STOP          0x08
```

Type definitions for lists and handles

```
/*********************************************************************
CAN message monitoring, list of CAN IDs not expected for receive
*********************************************************************/
typedef struct {
  COBID_TYPE  lst[Cc_CANIDLIST_LEN]; // list of CAN IDs
  UNSIGNED8   cur;                   // current number of entries in list
} Cc_CANIDLIST_HANDLE;


/*********************************************************************
Book section 5.4 "CANcrypt secure message table"
*********************************************************************/
typedef struct {
  COBID_TYPE  CAN_ID;        // CAN message ID of the secure message
  UNSIGNED8   EncryptFirst;  // first byte to encrypt
  UNSIGNED8   EncryptLen;    // number of bytes to encrypt
  UNSIGNED8   FunctMethod;   // function and methods
  UNSIGNED8   Producer;      // address (1-15) of the producer
} Cc_SEC_MSG_TABLE_ENTRY;


/*********************************************************************
For each entry in the secure message list, CANcrypt also needs a set of
variables to track processing of those.
*********************************************************************/
typedef struct {
  CAN_MSG     preamble;      // the preamble CAN message
  UNSIGNED16  time;          // timestamp of last message
  UNSIGNED8   count;         // message counter
  UNSIGNED8   pre_rx;        // TRUE, if preamble received
} Cc_SEC_MSG_TRACK_ENTRY;
```

Note: Internally used enumerations for state machines removed.

Call-back function prototypes

```
/*****************************************************************
DOES:    Call-back function, reports a system status change to the
         application
RETURNS: Nothing
*****************************************************************/
typedef void (*Cc_CB_EVENT) (
  UNSIGNED8    event,          // event that happened (Cc_EVENT_xxx)
  UNSIGNED32   param1,         // depending on the event, up to 3
  UNSIGNED32   param2,         // event related parameters
  UNSIGNED32   param3
);


/*****************************************************************
DOES:    Call-back function to pass on a CAN message to a buffer
RETURNS: TRUE, if message was accepted
         FALSE, if message could not be processed
*****************************************************************/
typedef UNSIGNED8 (*Cc_CAN_PUSH) (
  CAN_MSG    *pCAN            // CAN message to transfer
);


/*****************************************************************
DOES:    Call back function to read a generic parameter. Only called
         when request comes from a paired device and is secure.
RETURNS: Length of data in bytes, 0 for not available, else 1-4
*****************************************************************/
typedef UNSIGNED8 (*Cc_GA_READ) (
  UNSIGNED16   index,          // index to read
  UNSIGNED8    sub,            // subindex to read
  UNSIGNED8    dat[4]          // pointer to data buffer, copy
);                             // data to here


/*****************************************************************
DOES:    Call back function to write a generic parameter. Only called
         when request comes from apaired device and is secure.
RETURNS: Number of bytes written, 0 on error
*****************************************************************/
typedef UNSIGNED8 (*Cc_GA_WRITE) (
  UNSIGNED16   index,          // index to write
  UNSIGNED8    sub,            // subindex to write
  UNSIGNED8    length,         // length of data (1 - 4)
  UNSIGNED8    dat[4]          // data to write
);


/*****************************************************************
Main CANcrypt handle with all data required for CANcrypt operation
*****************************************************************/
typedef struct
{
For internal use only

} Cc_HANDLE;
#endif
/*---------------------- END OF FILE ----------------------*/
```

8.1.3 CANcrypt_api.h

This module contains the function definitions for actively controlling CANcrypt from an application..

```
/********************************************************************
MODULE:    CANcrypt_api.h, API functions
CONTAINS:  Definitions for CANcrypt Application Programming Interface
AUTHOR:    2017 Embedded Systems Academy, GmbH
HOME:      www.esacademy.com/cancrypt

    Licensed under the Apache License, Version 2.0 (the "License");
    you may not use this file except in compliance with the License.
    You may obtain a copy of the License at
    www.apache.org/licenses/LICENSE-2.0

    Unless required by applicable law or agreed to in writing, software
    distributed under the License is distributed on an "AS IS" BASIS,
    WITHOUT WARRANTIES OR CONDITIONS OF ANY KIND, either express or
    implied. See the License for the specific language governing
    permissions and limitations under the License.

VERSION:   0.10, 19-JAN-2017
********************************************************************/

#ifndef _CANCRYPT_API_H
#define _CANCRYPT_API_H

#include "CANcrypt_types.h"
```

CANcrypt initialization and controls

```
/********************************************************************
DOES:    Re-start of the CANcrypt system.
RETURNS: TRUE, if completed
         FALSE, if error in parameters passed
********************************************************************/
UNSIGNED8 Cc_Restart(
  Cc_HANDLE   *pCc,        // pointer to CANcrypt handle record
  UNSIGNED8    address,    // address of this device, set to zero if
                           // taken from config Ccnvol_GetGroupInfo()
  UNSIGNED32   control,    // Bit0-1: 00: No change to pairing
                           //         01: Restart pairing
                           //         10: Stop pairing
                           // Bit2-3: 00: No change to grouping
                           //         01: Restart grouping
                           //         10: Stop grouping
  // Call-back functions
  Cc_CB_EVENT fct_event,      // change of status, alert
  Cc_CAN_PUSH fct_pushTxFIFO, // put CAN message into Tx FIFO
  Cc_CAN_PUSH fct_pushTxNOW,  // transmit this CAN message now
  Cc_GA_READ  fct_readacc,    // generic read access
  Cc_GA_WRITE fct_writeacc    // generic write access
);
```

The *Cc_Restart()* function is the main initialization function for CANcrypt. The call-back functions *fct_event()* and *fct_pushTxFIFO()* are mandatory, as they are always used. The others may be set to NULL if not used.

```
/*****************************************************************
DOES:    Disconnect from the CANcrypt communication parners,
         sends a request to end pairing / grouping.
RETURNS: nothing
*****************************************************************/
void Cc_TxDisconnect(
  Cc_HANDLE *pCc,          // pointer to CANcrypt handle record
  UNSIGNED8 dest_addr,     // paired device ID (1-15) or 0 for group
  UNSIGNED8 reason         // reason for disconnecting, event/aler code
);


/*****************************************************************
DOES:    Generate a response message of type acknowledge or abort.
RETURNS: nothing
*****************************************************************/
void Cc_TxAckAbort(
  Cc_HANDLE *pCc,          // pointer to CANcrypt handle record
  UNSIGNED8 ack,           // TRUE for Ack, FALSE for Abort
  UNSIGNED8 dest_addr,     // destination device ID (1-15)
                           // or 0 for broadcast
  UNSIGNED8 key_id,        // key id for this acknowledge, 0 if unused
  UNSIGNED8 key_len        // key len for this acknowledge, 0 if unused
);


/*****************************************************************
DOES:    Generate an alert message.
RETURNS: nothing
*****************************************************************/
void Cc_TxAlert(
  Cc_HANDLE *pCc,          // pointer to CANcrypt handle record
  UNSIGNED8 dest_addr,     // destination device ID (1-15)
                           // or 0 for broadcast
  UNSIGNED16 alert         // 16bit alert or error code
);


/*****************************************************************
DOES:    Generate an identify message.
RETURNS: nothing
*****************************************************************/
void Cc_TxIdentify(
  Cc_HANDLE *pCc,          // pointer to CANcrypt handle record
  UNSIGNED16 version,      // CANcrypt version
  UNSIGNED8 key_id,        // key id desired
  UNSIGNED8 key_len,       // key len desired
  UNSIGNED8 cc_method,     // method desired (Cc_SECFCT_xxx)
  UNSIGNED8 cc_timing      // timing desired (Cc_TIMING_xxx)
);
```

Unexpected message monitoring

```
/*******************************************************************
          CAN receive monitoring for unexpected messages
DOES:     Init (erase) the list of CAN IDs
RETURNS:  TRUE, if list was erased
          FALSE, if list could not be erased
*******************************************************************/
UNSIGNED8 Cc_Init_IDList (
  Cc_CANIDLIST_HANDLE *pIDs      // pointer to list and length
);


/*******************************************************************
          CAN receive monitoring for unexpected messages
DOES:     Adds a CAN ID to the maintained list of IDs.
RETURNS:  TRUE, if ID was added
          FALSE, if ID could not be added
*******************************************************************/
UNSIGNED8 Cc_Addto_IDList(
  Cc_CANIDLIST_HANDLE *pIDs,     // pointer to list and length
  COBID_TYPE can_id              // CAN ID to add
);


/*******************************************************************
          CAN receive monitoring for unexpected messages
DOES:     Checks if a CAN ID is in the maintained list of IDs.
RETURNS:  TRUE, if ID is in list
          FALSE, if ID is NOT in list or error
*******************************************************************/
UNSIGNED8 Cc_Isin_IDList(
  Cc_CANIDLIST_HANDLE *pIDs,     // pointer to list and length
  COBID_TYPE can_id              // CAN ID to check
);
```

Secure message handling

```
/*******************************************************************
DOES:     Installs the secure message handlers by passing the
          appropriate secure message tables for transmit and receive.
RETURNS:  TRUE, if completed
          FALSE, if error in parameters passed
*******************************************************************/
UNSIGNED8 Cc_Load_Sec_Msg_Table(
  Cc_HANDLE *pCc,          // pointer to CANcrypt handle record
  Cc_SEC_MSG_TABLE_ENTRY *pMsgTblRx, // secure messages to receive
  Cc_SEC_MSG_TRACK_ENTRY *pMsgTrkRx, // variables for above
  Cc_SEC_MSG_TABLE_ENTRY *pMsgTblTx, // secure messages to receive
  UNSIGNED8              *pMsgTrkTcnt // counter for above
);
```

```
/********************************************************************
DOES:    Initialize the transmit and receive counters for the secure
         messages, may only be called "synchronized" for all
         paired / grouped devices, e.g. directly with pairing /
         grouping confirmation.
RETURNS: nothing
********************************************************************/
void Cc_Init_Sec_Msg_Table_Counter(
  Cc_HANDLE *pCc        // pointer to CANcrypt handle record
);


/********************************************************************
DOES:    This is the secure CAN transmit function of CANcrypt. It
         must be called before the transmit message is copied
         to the transmit FIFO, as a preamble might need to be
         inserted first.
RETURNS: 0: message ignored (not handled) by CANcrypt    ADD TO FIFO
         1: message is secured by CANcrypt,  ADD PREAMBLE&MSG TO FIFO
         2: message requires security, but we are not paired
                                                 DO NOT ADD TO FIFO
********************************************************************/
UNSIGNED8 Cc_Process_secMsg_Tx(
  Cc_HANDLE *pCc,         // pointer to CANcrypt handle record
  CAN_MSG *preamble,      // pointer to CAN buffer for preamble
  CAN_MSG *pCANtx         // pointer to CAN message to transmit
);


/********************************************************************
DOES:    This is the secure CAN Rx function of CANcrypt, needs to be
         called before a message received goes into receive FIFO.
RETURNS: 0: message ignored by CANcrypt,                  ADD TO FIFO
         1, message is a preamble,             DO NOT ADD TO FIFO
         2, secure message, and it is secure,          ADD TO FIFO
         3, message requires security, but we are not paired
                                               DO NOT ADD TO FIFO
********************************************************************/
UNSIGNED8 Cc_Process_secMsg_Rx(
  Cc_HANDLE *pCc,         // pointer to CANcrypt handle record
  CAN_MSG *pCANrx         // pointer to CAN message received
);
```

Time triggered processes

```
/********************************************************************
DOES:    Main CANcrypt householding functionality. Call cyclicly.
         Primarily monitors timeouts and satet transitions.
RETURNS: TRUE, if there was something to process
         FALSE, if there was nothing to do
********************************************************************/
UNSIGNED8 Cc_Process_Tick(
  Cc_HANDLE *pCc         // pointer to CANcrypt handle record
);
```

```
/***************************************************************
Same as above, but for individual tasks within CANcrypt:
bit and key generation process, pairing, grouping, monitoring
***************************************************************/
UNSIGNED8 Cc_Process_Key_Tick(Cc_HANDLE *pCc);
UNSIGNED8 Cc_Process_Pair_Tick(Cc_HANDLE *pCc);
UNSIGNED8 Cc_Process_Group_Tick(Cc_HANDLE *pCc);
UNSIGNED8 Cc_Process_Monitor_Tick(Cc_HANDLE *pCc);
```

The default implementation of *Cc_Process_Tick()* is shown below.

```
/***************************************************************
DOES:    Main CANcrypt householding functionality. Call cyclicly.
         Primarily monitors timeouts and satet transitions.
RETURNS: TRUE, if there was something to process
         FALSE, if there was nothing to do
***************************************************************/
UNSIGNED8 Cc_Process_Tick(
  Cc_HANDLE *pCc         // pointer to a handle record
)
{
UNSIGNED8 ret_val = FALSE;

#ifdef Cc_USE_MONITORING
  // did we receive an unexpected message?
  ret_val = Cc_Process_Monitor_Tick(pCc);
#endif

#ifdef Cc_USE_SECURE_MSG
  if (! ret_val)
  { // secure message handling OK?
    ret_val = Cc_Process_secMsg_Tick(pCc);
  }
#endif

#ifdef Cc_USE_PAIRING
  if (! ret_val)
  { // work on key generation timings and timeouts
    ret_val = Cc_Process_Key_Tick(pCc);
    if (! ret_val)
    { // work on pairing timings and timeouts
      ret_val = Cc_Process_Pair_Tick(pCc);
    }
  }
#endif

#ifdef Cc_USE_GROUPING
  if (! ret_val)
  { // work on grouping timings and timeouts
    ret_val = Cc_Process_Group_Tick(pCc);
  }
#endif

  return ret_val;
}
```

The function *Cc_Process_Tick()* needs to be called cyclically, preferably multiple times per millisecond. If called less frequent or integrated in a Real-Time Operating System, the call should be

```
while (Cc_Process_Tick(pCc)
{
}
```

This ensures that it keeps processing as long as there are some CANcrypt tasks to execute.

CAN receive triggered processes

```
/****************************************************************
DOES:    This is the main CAN receive function of CANcrypt, must be
         called directly from CAN receive interrupt. Distributes a
         message to the other Cc_Process_xxx_Rx() functions.
RETURNS: TRUE, if this message was processed by CANcrypt
         FALSE, if this message was ignored by CANcrypt
****************************************************************/
UNSIGNED8 Cc_Process_Rx(
  Cc_HANDLE *pCc,        // pointer to CANcrypt handle record
  CAN_MSG *pCANrx        // pointer to CAN message received
);


/****************************************************************
Same as above, but for individual tasks within CANcrypt:
bit and key generation process, pairing, grouping
****************************************************************/
UNSIGNED8 Cc_Process_Key_Rx(Cc_HANDLE *pCc, CAN_MSG *pCANrx);
UNSIGNED8 Cc_Process_Pair_Rx(Cc_HANDLE *pCc, CAN_MSG *pCANrx);
UNSIGNED8 Cc_Process_Group_Rx(Cc_HANDLE *pCc, CAN_MSG *pCANrx);
UNSIGNED8 Cc_Process_Monitor_Rx(Cc_HANDLE *pCc, CAN_MSG *pCANrx);

#endif
/*--------------------- END OF FILE ---------------------------*/
```

The default implementation for *Cc_Process_Rx()* is as follows:

```
/****************************************************************
DOES:    This is the main CAN receive function of CANcrypt, must be
         called directly from CAN receive interrupt. Distributes a
         message to the other Cc_Process_xxx_Rx() functions.
RETURNS: TRUE, if this message was processed by CANcrypt
         FALSE, if this message was ignored by CANcrypt
****************************************************************/
UNSIGNED8 Cc_Process_Rx(
  Cc_HANDLE *pCc,        // pointer to CANcrypt handle record
  CAN_MSG *pCANrx        // pointer to CAN message received
)
```

```
{
UNSIGNED8 ret_val = FALSE;

#ifdef Cc_USE_MONITORING
  // is message in list of unexpected messages?
  ret_val = Cc_Process_Monitor_Rx(pCc,pCANrx);
#endif

#ifdef Cc_USE_PAIRING
  if (! ret_val)
  { // is message for key generation?
    ret_val = Cc_Process_Key_Rx(pCc,pCANrx);
    if (! ret_val)
    { // is message for pairing?
      ret_val =  Cc_Process_Pair_Rx(pCc,pCANrx);
    }
  }
#endif

#ifdef Cc_USE_GROUPING
  if (! ret_val)
  { // is message for grouping?
    ret_val =  Cc_Process_Group_Rx(pCc,pCANrx);
  }
#endif

#ifdef Cc_USE_SECURE_MSG
  if (! ret_val)
  { // is message a secure message?
    ret_val = Cc_Process_secMsg_Rx(pCc,pCANrx);
    if ((ret_val & 1) == 1)
    { // message was processed, do not add to FIFO
      ret_val = TRUE;
    }
    else
    {
      ret_val = FALSE;
    }
  }
#endif

  return ret_val;
}
```

The *Cc_Process_Rx()* function needs to be called directly from the CAN receive interrupt.

The first three calls check if this CAN message is a CANcrypt command, status, event or bit claiming message used by the separate CANcrypt tasks. The call to *Cc_Process_secMsg_Rx()* verifies if this is a secured message.

At the end, TRUE is returned, if this message should not be passed on to the application. The driver/interrupt has to dismiss/ignore it.

8.2 CANcrypt_userfct.h

The definitions and functions of this file have been described in chapter 6 CANcrypt customizable functions. For the full C source code listing of the default implementation, see Appendix B.

8.3 Low-level driver interfacing

CANcrypt requires access to the following system resources:

1.) The CAN communication interface
2.) A one Millisecond timer
3.) Non-volatile memory for storage of configuration and keys

This chapter describes the detailed requirements for these interfaces.

8.3.1 CAN interface access

To simplify required code changes to existing implementations, the programming interface provides hooks to typical CAN driver processing functions. Several functions that are time critical typically need to be integrated at the lowest driver level, directly in the CAN receive interrupt routine. When integrated at this level, the changes to the user or application level are minimal.

The diagram below illustrates a typical CAN driver with FIFOs (First-In, First-Out buffers). A CAN interrupt service routine copies received CAN messages into a receive FIFO. The messages are processed later by a protocol stack (such as CANopen) or directly by the application. Messages transmitted by the application or protocol stack are added to a transmit FIFO and from there go into a CAN controller transmit buffer.

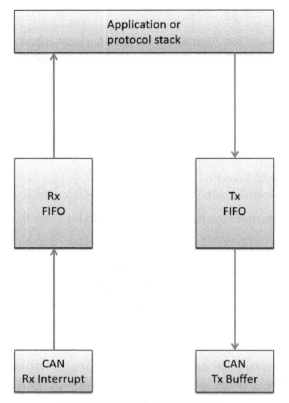

TYPICAL FIFO CAN DRIVER

CANcrypt functions can be fully integrated into this scheme, minimizing the impact on the application or the protocol stack.

Once the CANcrypt system is integrated and active (paired device found), no changes are required in regard to receiving CAN messages. The CANcrypt receive handler (CANcrypt process Rx) is integrated into the CAN receive interrupt handler and copies received secure messages only after they have been authenticated and decrypted.

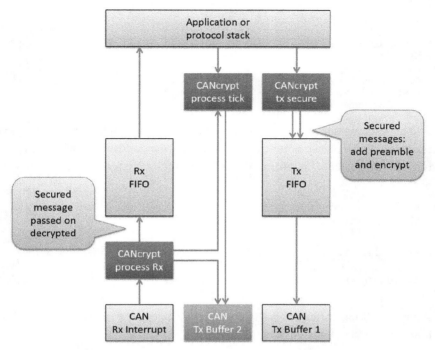

FUNCTION "HOOKS" BETWEEN CANCRYPT AND DRIVER

In regard to transmission of secure messages, the application or protocol stack can add unsecured messages to the transmit FIFOs in the same manner as without CANcrypt. However, for a secure transmission, CANcrypt generates the appropriate preamble and encrypted message.

If used with the option for the fastest bit-generation cycle (direct response to trigger, not random delay), a dedicated CAN transmit buffer (CAN Tx buffer 2 in figure above) is required.

The background handler (process tick) manages the pairing of devices and updating the dynamic key.

Moving CAN messages

In CANcrypt a CAN message is defined as a structure of the CAN ID (data type of 16 bits or 32 bits depending if CAN is used with 11-bit or 29-bit CAN message identifiers), the data length and the data. This is a common definition used by many drivers. Sometimes an additional time and or control filed is added. Here we use the minimal version as shown below.

```
/*****************************************************************
Data structure for a single CAN message
*****************************************************************/
typedef struct
{ // order optimized for allignment
  UNSIGNED8 BUF[8];              // Data buffer
  COBID_TYPE ID;                 // Message Identifier
  UNSIGNED8 LEN;                 // Data length (0-8)
} CAN_MSG;
```

When a CAN message is passed on to a FIFO or another handler, then the parameters passed are only a pointer to the CAN message structure and the value returned is a Boolean. The return value is TRUE, if the message was passed on without errors.

```
/*****************************************************************
DOES:    Function to pass on a CAN message to a buffer
RETURNS: TRUE, if message was accepted
         FALSE, if message could not be processed
*****************************************************************/
typedef UNSIGNED8 (*Cc_CAN_PUSH) (
  CAN_MSG *pCAN                   // CAN message to transfer
);
```

When initializing CANcrypt via the *Cc_Restart()* function, a total of two such driver functions need to be passed on to CANcrypt. These are:

fct_pushTxFIFO
This function places the CAN message passed into the regular transmit FIFO/queue. If there are already messages in this transmit FIFO it will go out after all the messages in the FIFO went out.

fct_pushTxNOW
This function bypasses the transmit FIFO and directly places this CAN message into a CAN transmit buffer of the CAN controller. This function is only needed, if the direct bit-claiming mode is used instead of the random delay method.

```
/*****************************************************************
MODULE:    CANcrypt_can.h, CAN definitions and functions
CONTAINS:  CAN interface definitions for the CANcrypt system
AUTHOR:    2017 Embedded Systems Academy, GmbH
HOME:      www.esacademy.com/cancrypt

  Licensed under the Apache License, Version 2.0 (the "License");
  you may not use this file except in compliance with the License.
  You may obtain a copy of the License at
  www.apache.org/licenses/LICENSE-2.0
```

```
VERSION:   0.10, 19-JAN-2017
****************************************************************/

#ifndef _CANCRYPT_CAN_H
#define _CANCRYPT_CAN_H

#include "Cc_user_types.h"


/*****************************************************************
Data structure for a single CAN message
*****************************************************************/
typedef struct
{ // order optimized for allignment
  UNSIGNED8 BUF[8];                // Data buffer
  COBID_TYPE ID;                   // Message Identifier
  UNSIGNED8 LEN;                   // Data length (0-8)
} CAN_MSG;


/*****************************************************************
DOES:    This function initializes the CAN interface.
RETURNS: 1 if init is completed
         0 if init failed
*****************************************************************/
UNSIGNED8 CCHW_Init (
  UNSIGNED16 BaudRate    // 1000, 800, 500, 250, 125 or 50
  );


/*****************************************************************
DOES:    This function implements a CAN transmit FIFO. With each
         function call a message is added to the FIFO.
RETURNS: 1 Message was added to the transmit FIFO
         0 If FIFO is full, message was not added
*****************************************************************/
UNSIGNED8 CCHW_PushMessage (
  CAN_MSG MEM_FAR *pTransmitBuf // CAN message to send
  );


/*****************************************************************
DOES:    This function retrieves a CAN message from the receive FIFO.
RETURNS: 1 Message was pulled from receive FIFO
         0 Queue empty, no message received
*****************************************************************/
UNSIGNED8 CCHW_PullMessage (
  CAN_MSG MEM_FAR *pReceiveBuf // Buffer for received message
  );
```

```
/******************************************************************
DOES:      Transmission of a CANcrypt high priority message.
NOTE:      Must be transmit immediately, no delay. Bypasses Tx FIFO.
           ONLY REQUIRED FOR DIRECT BIT CLAIMING MODE
RETURNS: Nothing
******************************************************************/
UNSIGNED8 CCHW_TransmitNow(
  CAN_MSG *pMsg
);


/******************************************************************
DOES:      Enable/Disable of CAN receive interrupt
RETURNS: Nothing
******************************************************************/
void CCHW_EnableCANrx (void);
void CCHW_DisableCANrx (void);


#endif
/*---------------------- END OF FILE ----------------------------*/
```

8.3.2 Random numbers, timer and timeout

CANcrypt uses timings based on multiple of milliseconds. Only two functions are required.

```
/******************************************************************
MODULE:    CANcrypt_hwsys.h, Misc HW system functions
CONTAINS:  Random number generation and timers
AUTHOR:    2017 Embedded Systems Academy, GmbH
HOME:      www.esacademy.com/cancrypt

  Licensed under the Apache License, Version 2.0 (the "License");
  you may not use this file except in compliance with the License.
  You may obtain a copy of the License at
  www.apache.org/licenses/LICENSE-2.0

  Unless required by applicable law or agreed to in writing, software
  distributed under the License is distributed on an "AS IS" BASIS,
  WITHOUT WARRANTIES OR CONDITIONS OF ANY KIND, either express or
  implied. See the License for the specific language governing
  permissions and limitations under the License.

VERSION:   0.10, 19-JAN-2017
******************************************************************/

#ifndef _CANCRYPT_HWSYS_H
#define _CANCRYPT_HWSYS_H


#include "Cc_user_types.h"
```

```
/********************************************************************
DOES:     Generates a random value.
NOTE:     Must be suitable for security use, shall not produce the
          same sequence of numbers on every reset or power up!!
RETURNS: Random value
********************************************************************/
INTEGER32 CCHW_Rand(
  void
);



/********************************************************************
DOES:     This function reads a 1 millisecond timer tick. The timer
          tick must be a UNSIGNED16 and must be incremented once per
          millisecond.
RETURNS: 1 millisecond timer tick
********************************************************************/
UNSIGNED16 CCHW_GetTime (
  void
  );



/********************************************************************
DOES:     This function compares a UNSIGNED16 timestamp to the
          internal timer tick and returns 1 if the timestamp
          expired/passed.
RETURNS: 1 if timestamp expired/passed
         0 if timestamp is not yet reached
NOTES:    The maximum timer runtime measurable is 0x7FFF
********************************************************************/
UNSIGNED8 CCHW_IsTimeExpired (
  UNSIGNED16 timestamp // Timestamp to be checked for expiration
  );


#endif
/*---------------------- END OF FILE ----------------------------*/
```

8.3.3 Non-Volatile memory access

Non-volatile memory, like EEPROM, is needed to store keys and configurations. If a system is hard-coded, these parameters could also all be stored in code memory area.

```
/*********************************************************************
MODULE:    CANcrypt_nvol.h, access to non volatile memory
CONTAINS:  Functions that access data stored in NVOL memory
AUTHOR:    2017 Embedded Systems Academy, GmbH
HOME:      www.esacademy.com/cancrypt

   Licensed under the Apache License, Version 2.0 (the "License");
   you may not use this file except in compliance with the License.
   You may obtain a copy of the License at
   www.apache.org/licenses/LICENSE-2.0

   Unless required by applicable law or agreed to in writing, software
   distributed under the License is distributed on an "AS IS" BASIS,
   WITHOUT WARRANTIES OR CONDITIONS OF ANY KIND, either express or
   implied. See the License for the specific language governing
   permissions and limitations under the License.

VERSION:   0.10, 19-JAN-2017
*********************************************************************/


#ifndef _CANCRYPT_NVOL_H
#define _CANCRYPT_NVOL_H

#include "CANcrypt_types.h"
```

Grouping information

When grouping is used, devices need to know which address/device number they have and which communication partners are expected. CANcrypt configuration supports a list of mandatory and a list of optional partners. Grouping initialization ends when all mandatory partners were found.

```
/*********************************************************************
DOES:    This function retrieves the last saved grouping parameters
RETURNS: Both my_addr and grp_info
*********************************************************************/
void Ccnvol_GetGroupInfo(
  UNSIGNED8 *my_addr,    // address (1-15) to use by this device
  UNSIGNED16 *grp_mand,  // Mandatory devices to start up
                         // bits 1-15 set for each device in group
                         // bit 0 set if grouping is disabled
  UNSIGNED16 *grp_opt    // Optional devices to start up
                         // bits 1-15 set for each device in group
                         // bit 0 set if grouping is disabled
);
```

```
/***************************************************************
DOES:    This function saves the current grouping parameters
RETURNS: TRUE, if saved, FALSE if failed
***************************************************************/
UNSIGNED8 Ccnvol_SaveGroupInfo(
  UNSIGNED8 my_addr,     // address (1-15) to use by this device
  UNSIGNED16 grp_info    // bits 1-15 set for each device in group
                         // bit 0 set if grouping is disabled
);
```

Key hierarchy access

These functions provide access to the key hierarchy. Erase or save commands are only called by CANcrypt, if a request with the appropriate authorization has been received.

```
/***************************************************************
BOOK:    Section 6.1 "Key hierarchy access"
DOES:    This function directly returns a key from the key hierarchy
RETURNS: Pointer to the key or NULL if not available
***************************************************************/
UNSIGNED32 *Ccnvol_GetPermKey(
  UNSIGNED8 key_ID       // key major ID, 2 to 6
);


/***************************************************************
BOOK:    Section 6.1 "Key hierarchy access"
DOES:    This function erases a key from the key hierarchy. Will only
         be called from CANcrypt, if called from authorized
         configurator.
RETURNS: TRUE, if key was erased, else FALSE
***************************************************************/
UNSIGNED8 *Ccnvol_ErasePermKey(
  UNSIGNED8 key_ID       // key major ID, 2 to 6
);


/***************************************************************
BOOK:    Section 6.1 "Key hierarchy access"
DOES:    This function saves a key to the key hierarchy. Will only
         be called from CANcrypt, if called from authorized
         configurator.
RETURNS: TRUE, if key was erased, else FALSE
***************************************************************/
UNSIGNED8 *Ccnvol_SavePermKey(
  UNSIGNED8 key_ID,      // key major ID, 2 to 6
  UNSIGNED32 *pkey       // pointer to key data
);

#endif
/*--------------------- END OF FILE ----------------------*/
```

8.4   Secure message configuration

The secure message list record is defined in CANcrypt_types.h.

```
/*********************************************************************
Book section 5.4 "CANcrypt secure message table"
*********************************************************************/
typedef struct {
  COBID_TYPE  CAN_ID;       // CAN message ID of the secure message
  UNSIGNED8   EncryptFirst; // first byte to encrypt
  UNSIGNED8   EncryptLen;   // number of bytes to encrypt
  UNSIGNED8   FunctMethod;  // function and methods
  UNSIGNED8   Producer;     // address (1-15) of the producer
} Cc_SEC_MSG_TABLE_ENTRY;
```

The application needs to provide two arrays with these records, one array with the secure messages to receive, and one with the secure messages to transmit. The last entry in the list needs to be all values FFh to indicate the end of the table.

The example below shows lists with two secure transmit messages and three secure receive messages.

```
// secure message table for transmit
Cc_SEC_MSG_TABLE_ENTRY TxSecMg[3] = {
  0x0183, 2, 4, 0, 3,
  0x0283, 0, 6, 0, 3,
  0xFFFF, 0xFF, 0xFF, 0xFF, 0xFF
};

// secure message table for receive
Cc_SEC_MSG_TABLE_ENTRY RxSecMg[4] = {
  0x0181, 2, 4, 0, 1,
  0x0182, 2, 4, 0, 1,
  0x0281, 0, 6, 0, 1,
  0xFFFF, 0xFF, 0xFF, 0xFF, 0xFF
};
```

When passing these lists on to CANcrypt via the *Cc_Load_Sec_Msg_Table()* function, CANcrypt also requires lists with state tracking information. For the example above they can be allocated as follows:

```
// secure message tracking info required by CANcrypt
UNSIGNED8 TxTrk[2];
Cc_SEC_MSG_TRACK_ENTRY RxTrk[3];
```

The call to activate the lists above is:

```
Cc_Load_Sec_Msg_Table(&CcH,RxSecMg,RxTrk,TxSecMg,TxTrk);
```

8.5   Demo and driver example

The CAN blocks on various microcontrollers may be quite different. This section shows implementation details for the NXP LPC1768 microcontroller.

8.5.1   CAN queue / FIFO

Often a CAN software FIFO buffer (first-in-first-out) queue is used as a hardware abstraction layer. If an application uses the same FIFO on different target micro-controllers, than it can be independent from the various differences of the CAN controllers. This sections shows the simple implementation of a CAN receive and transmit FIFO.

There are separate FIFOs for transmit and receive. The functions for each FIFO are a flush (erase), getting the current "in" or "out" pointers and a "done" call when data was copied. See the next chapter for a usage example.

```
/********************************************************************
MODULE:     canfifo.c, CAN FIFO demo - message queues
CONTAINS:   CAN transmit and receive FIFO
COPYRIGHT:  Embedded Systems Academy GmbH, 2016-2017
HOME:       www.esacademy.com/cancrypt
LICENSE:    FOR EDUCATIONAL AND EVALUATION PURPOSE ONLY!
CONTACT:    Contact info@esacademy.de for other available licenses

Unless required by applicable law or agreed to in writing, software
distributed under the License is distributed on an "AS IS" BASIS,
WITHOUT WARRANTIES OR CONDITIONS OF ANY KIND, either express or
implied.

VERSION:    0.10, 19-JAN-2017
********************************************************************/

#include "CANcrypt_includes.h"


#if (TXFIFOSIZE != 0) && (TXFIFOSIZE != 4) && (TXFIFOSIZE != 8) &&
(TXFIFOSIZE != 16) && (TXFIFOSIZE != 32) && (TXFIFOSIZE != 64)
 #error "TXFIFOSIZE must be 0 (deactivated), 4, 8, 16, 32 or 64"
#endif
#if (RXFIFOSIZE != 0) && (RXFIFOSIZE != 4) && (RXFIFOSIZE != 8) &&
(RXFIFOSIZE != 16) && (RXFIFOSIZE != 32) && (RXFIFOSIZE != 64)
 #error "RXFIFOSIZE must be 0 (deactivated), 4, 8, 16, 32 or 64"
#endif


/********************************************************************
MODULE VARIABLES
********************************************************************/
```

```
typedef struct
{
#if (TXFIFOSIZE > 0)
  CAN_MSG TxFifo[TXFIFOSIZE];
#endif
#if (RXFIFOSIZE > 0)
  CAN_MSG RxFifo[RXFIFOSIZE];
#endif
#if (TXFIFOSIZE > 0)
  UNSIGNED8 TxIn;
  UNSIGNED8 TxOut;
#endif
#if (RXFIFOSIZE > 0)
  UNSIGNED8 RxIn;
  UNSIGNED8 RxOut;
#endif
} CANFIFOINFO;

// Module variable with all FIFO information
CANFIFOINFO mCF;


/*********************************************************************
PUBLIC FUNCTIONS
*********************************************************************/
```

Transmit FIFO / queue

```
#if (TXFIFOSIZE > 0)
/*********************************************************************
DOES:    Flushes / clears the TXFIFO, all data stored in FIFO is lost
RETURNS: nothing
*********************************************************************/
void CANTXFIFO_Flush (
  void
  )
{
  mCF.TxIn = 0;
  mCF.TxOut = 0;
}


/*********************************************************************
DOES:    Returns a CAN message pointer to the next free location in
         the FIFO. The application may then copy a CAN message to the
         location given by the pointer and MUST call
         CANTXFIFO_InDone() when copy completed.
RETURNS: CAN message pointer into FIFO
         NULL if FIFO is full
*********************************************************************/
```

```
CAN_MSG *CANTXFIFO_GetInPtr (
  void
  )
{
UNSIGNED8 ovr; // check if FIFO is full

  ovr = mCF.TxIn + 1;
  ovr &= (TXFIFOSIZE-1);

  if (ovr != mCF.TxOut)
  {// FIFO is not full
    return &(mCF.TxFifo[mCF.TxIn]);
  }
  return 0;
}

/*******************************************************************
DOES:    Must be called by the application after data was copied into
         the FIFO, this increments the internal IN pointer to the
         next free location in the FIFO.
RETURNS: nothing
*******************************************************************/
void CANTXFIFO_InDone (
  void
  )
{
  // Increment IN pointer
  mCF.TxIn++;
  mCF.TxIn &= (TXFIFOSIZE-1);
}

/*******************************************************************
DOES:    Returns a CAN message pointer to the next OUT message in the
         FIFO. The application may then copy the CAN message from the
         location given by the pointer to the desired destination and
         MUST call CANTXFIFO_OutDone() when done.
RETURNS: CAN message pointer into FIFO
         NULL if FIFO is empty
*******************************************************************/
CAN_MSG *CANTXFIFO_GetOutPtr (
  void
  )
{
  if (mCF.TxIn != mCF.TxOut)
  { // message available in FIFO
    return &(mCF.TxFifo[mCF.TxOut]);
  }
  return 0;
}

/*******************************************************************
DOES:    Must be called by application after data was copied from the
         FIFO, this increments the internal OUT pointer to the next
         location in the FIFO.
RETURNS: nothing
*******************************************************************/
```

```
void CANTXFIFO_OutDone (

  void
  )
{
  mCF.TxOut++;
  mCF.TxOut &= (TXFIFOSIZE-1);
}
#endif // (TXFIFOSIZE > 0)
```

Receive FIFO / queue

```
#if (RXFIFOSIZE > 0)
/*********************************************************************
DOES:    Flushes / clears the RXFIFO, all data stored in FIFO is lost
RETRUNS: nothing
*********************************************************************/
void CANRXFIFO_Flush (
  void
  )
{
  mCF.RxIn = 0;
  mCF.RxOut = 0;
}

/*********************************************************************
DOES:    Returns a CAN message pointer to the next free location in
         the FIFO. The application may then copy a CAN message to the
         location given by the pointer and MUST call
         CANRXFIFO_InDone() when copy completed.
RETURNS: CAN message pointer into FIFO, NULL if FIFO is full
*********************************************************************/
CAN_MSG *CANRXFIFO_GetInPtr (
  void
  )
{
UNSIGNED8 ovr; // check if FIFO is full

  ovr = mCF.RxIn + 1;
  ovr &= (RXFIFOSIZE-1);

  if (ovr != mCF.RxOut)
  {// FIFO is not full
    return &(mCF.RxFifo[mCF.RxIn]);
  }
  return 0;
}

/*********************************************************************
DOES:    Must be called by the application after the data was copied
         into the FIFO, this increments the internal IN pointer to
         the next free location in the FIFO.
RETURNS: nothing
*********************************************************************/
```

```
void CANRXFIFO_InDone (
  void
  )
{
  // Increment IN pointer
  mCF.RxIn++;
  mCF.RxIn &= (RXFIFOSIZE-1);
}


/*******************************************************************
DOES:    Returns a CAN message pointer to the next OUT message in the
         FIFO. The application may then copy the CAN message from the
         location given by the pointer to the desired destination and
         MUST call CANRXFIFO_OutDone() when done.
RETURNS: CAN message pointer into FIFO, NULL if FIFO is empty
*******************************************************************/
CAN_MSG *CANRXFIFO_GetOutPtr (
  void
  )
{
  if (mCF.RxIn != mCF.RxOut)
  { // message available in FIFO
    return &(mCF.RxFifo[mCF.RxOut]);
  }
  return 0;
}


/*******************************************************************
DOES:    Must be called by application after data was copied from the FIFO,
         this increments the internal OUT pointer to the next location
         in the FIFO.
RETURNS: nothing
*******************************************************************/
void CANRXFIFO_OutDone (
  void
  )
{
  mCF.RxOut++;
  mCF.RxOut &= (RXFIFOSIZE-1);
}
#endif // (RXFIFOSIZE > 0)

/*----------------------- END OF FILE -----------------------*/
```

8.5.2 NXP LPC17xx driver

In the CANcrypt context a CAN driver provides functions to place messages into a FIFO or to retrieve it. In addition, the CANcrypt specific transmit function to by-pass the transmit FIFO must be provided, if the direct response mode is used for the bit-claiming cycles.

```
/*********************************************************************
MODULE:     MCOHW_LPC1, CAN Driver Demo
CONTAINS:   CAN Driver implementation for NXP LPC17xx derivatives with
            CAN interface. Compiled and Tested with ARM/Keil Tools
COPYRIGHT:  Embedded Systems Academy GmbH, 2016-2017
HOME:       www.esacademy.com/cancrypt
LICENSE:    FOR EDUCATIONAL AND EVALUATION PURPOSE ONLY!
CONTACT:    Contact info@esacademy.de for other available licenses

Unless required by applicable law or agreed to in writing, software
distributed under the License is distributed on an "AS IS" BASIS,
WITHOUT WARRANTIES OR CONDITIONS OF ANY KIND, either express or
implied.

VERSION:    0.10, 19-JAN-2017
*********************************************************************/

#include "CANcrypt_includes.h"
#include <string.h>
#include "LPC17xx.h" // LPC17xx Peripheral Registers

// Common CAN bit rates BTR values for 12MHz CAN Clock
#define    CANBitrate20k_12MHz       0x003E401DUL
#define    CANBitrate50k_12MHz       0x002B400EUL
#define    CANBitrate125k_12MHz      0x002B4005UL
#define    CANBitrate250k_12MHz      0x002B4002UL
#define    CANBitrate500k_12MHz      0x00274001UL
#define    CANBitrate800k_12MHz      0x00394000UL
#define    CANBitrate1000k_12MHz     0x00364000UL

// declare interrupts, for usage in IRQ Vector
void __irq CAN_IRQHandler (void);
void __irq TIMER0_IRQHandler (void);


/*********************************************************************
GLOBAL VARIABLES
*********************************************************************/

// global timer/conter variable, incremented every millisecond
UNSIGNED16 volatile gTimCnt = 0;

// handle for CANcrypt
extern Cc_HANDLE CcH;

// seed for random generation, located in unused RAM
INTEGER32 *pPseudoRand = (INTEGER32 *) 0x10006FF0;

// permanent key used, here just a demo default value
// CHANGE TO OWN KEY WHEN USING !!!
UNSIGNED32 perm_key[Cc_PERMKEY_LEN32] = {
  0x33339999,0x8888CCCC,0xEEEEBBBB,0x5555AAAA
  };
```

```
/*****************************************************************
LOCAL VARIABLES
*****************************************************************/
// current CAN message received
CAN_MSG CANrx;


/*****************************************************************
LOCAL FUNCTIONS
*****************************************************************/
static void MCOHW_CAN1ISR_Rx (void);
static void MCOHW_CAN1ISR_Err (void);


/*****************************************************************
PUBLIC FUNCTIONS
*****************************************************************/
```

Get random number

```
/*****************************************************************
DOES:    Generates a random value.
NOTE:    Must be suitable for security use, shall not produce the
         same sequence of numbers on every reset or power up!!
RETURNS: Random value
*****************************************************************/
INTEGER32 CCHW_Rand(void)
{ // use from random generator AND HW maintained value
  return rand() + *pPseudoRand;
}
```

Get NVOL stored configurations, keys

```
#ifdef Cc_USE_GROUPING
/*****************************************************************
DOES:    This function retrieves the last saved grouping parameters
RETURNS: Both my_addr and grp_info
*****************************************************************/
void Ccnvol_GetGroupInfo(
  UNSIGNED8 *my_addr,   // address (1-15) to use by this device
  UNSIGNED16 *grp_mand, // Mandatory devices to start up
                        // bits 1-15 set for each device in group
                        // bit 0 set if grouping is disabled
  UNSIGNED16 *grp_opt   // Optional devices to start up
                        // bits 1-15 set for each device in group
                        // bit 0 set if grouping is disabled
)
{ // here hard-coded combination
  *my_addr  = NODEID;
  *grp_mand = 0x008C;   // nodes 2 & 3 & 7
  *grp_opt  = 0x808C;   // nodes 2 & 3 & 7 & 15
}
#endif
```

```
/*********************************************************************
BOOK:     Section 6.1 "Key hierarchy access"
DOES:     This function directly returns a key from the key hierarchy
RETURNS: Pointer to the key or NULL if not available
*********************************************************************/
UNSIGNED32 *Ccnvol_GetPermKey(
  UNSIGNED8 key_ID      // key major ID, 2 to 6
)
{
  return &(perm_key[0]);
}
```

Lower level CAN functions

```
/*********************************************************************
DOES:     Enable/Disable of CAN receive interrupt
RETURNS: Nothing
*********************************************************************/
void CCHW_EnableCANrx (void)
{
  LPC_CAN1->IER = 0x000000AD;
}

void CCHW_DisableCANrx (void)
{
  LPC_CAN1->IER = 0x00000000;
  LPC_CAN1->IER = 0x00000000; // second covers pipeline side effects
}


/*********************************************************************
DOES:     This function implements the initialization of the CAN interface.
RETURNS: 1 if init is completed
         0 if init failed, bit INIT of MCOHW_GetStatus stays 0
*********************************************************************/
UNSIGNED8 CCHW_Init (
  UNSIGNED16 baudrate
  )
{
UNSIGNED32 *pAddr;
UNSIGNED32 p;
UNSIGNED32 btr;

  switch (baudrate)
  {
    case 20:
      btr = CANBitrate20k_12MHz;
      break;

    case 50:
      btr = CANBitrate50k_12MHz;
      break;
```

```
    case 125:
      btr = CANBitrate125k_12MHz;
      break;
    case 250:
      btr = CANBitrate250k_12MHz;
      break;

    case 500:
      btr = CANBitrate500k_12MHz;
      break;

    case 800:
      btr = CANBitrate800k_12MHz;
      break;

    case 1000:
      btr = CANBitrate1000k_12MHz;
      break;

    default:
      return 0;  // Not supported
}

// Init Tx FIFO
CANTXFIFO_Flush();

// Init RxFIFO
CANRXFIFO_Flush();

// Enable clock to the peripheral
LPC_SC->PCONP |= (1 << 13);
// Enable Pins for CAN port 1
LPC_PINCON->PINSEL0 |= (1 << 0); // Pin P0.0 used as RD1 (CAN1)
LPC_PINCON->PINSEL0 |= (1 << 2); // Pin P0.1 used as TD1 (CAN1)

// Reset and disable all message filters
// Acceptance Filter Mode Register = off !
LPC_CANAF->AFMR = 0x00000001L;

LPC_CAN1->MOD  = 1;       // Go into Reset mode
LPC_CAN1->IER  = 0;       // Disable All Interrupts
LPC_CAN1->GSR  = 0;       // Clear Status register
LPC_CAN1->CMR  = 0x0E;    // Clear receive buffer, overrun, abort tx
LPC_CAN1->BTR  = btr;     // Set bit timing

// Enter Normal Operating Mode
LPC_CAN1->MOD  = 0;       // Operating Mode

// Enable Receive Interrupt (bit 0), Error Warning (bit 2),
// Data overrun (bit 3), Error passive (bit 5), Bus error (bit 7)
LPC_CAN1->IER = 0x000000AD;

// Now work on Acceptance Filter Configuration
// Acceptance Filter Mode Register = off !
LPC_CANAF->AFMR = 0x00000001;
```

```
// Initialize pointer to filter RAM
pAddr = (UNSIGNED32 *) LPC_CANAF_RAM_BASE;
p = 0;


// Set pointer for Standard Frame Individual
// Standard Frame Explicit
LPC_CANAF->SFF_sa = p;

// Set pointer for Standard Frame Groups
// Standard Frame Group Start Address Register
LPC_CANAF->SFF_GRP_sa = p;

// Set pointer for Standard Frame Groups
// Receive all 11bit CAN IDS
*pAddr = (0x000 << 16) + 0x7FF;

p += 4;

// Set pointer for Extended Frame Individual
// Extended Frame Start Address Register
LPC_CANAF->EFF_sa = p;

// Set pointer for Extended Frame Groups
// Extended Frame Group Start Address Register
LPC_CANAF->EFF_GRP_sa = p;

// Set ENDofTable
// End of AF Tables Register
LPC_CANAF->ENDofTable = p;

// Acceptance Filter Mode Register, start using filter
LPC_CANAF->AFMR = 0x00000000;

// Initialize Timer Interrupt
LPC_TIM0->MR0 = 17999; // 1mSec = 18.000-1 counts
LPC_TIM0->MCR = 3;      // Interrupt and Reset on MR0
LPC_TIM0->TCR = 1;      // Timer0 Enable

// Init Interrupts
NVIC_EnableIRQ(TIMER0_IRQn);   // enable TIMER0 interrupt
NVIC_EnableIRQ(CAN_IRQn);   // enable CAN interrupt

// configure run and error LED outputs
LPC_GPIO1->FIODIR |= (1UL << 28) | (1UL << 29);

  return 1;
}
```

```
/****************************************************************
DOES:     This function checks the CAN receive queue / FIFO. With each
          function call a message is pulled from the queue.
RETURNS: 1 Message was pulled from receive queue
         0 Queue empty, no message received
****************************************************************/
UNSIGNED8 CCHW_PullMessage (
  CAN_MSG MEM_FAR *pReceiveBuf
  )
{
CAN_MSG MEM_BUF *pSrc;


  // Check if message is in Rx FIFO
  pSrc = CANRXFIFO_GetOutPtr();
  if (pSrc != 0)
  {
    // copy message
    memcpy(pReceiveBuf,pSrc,sizeof(CAN_MSG));
    // copying complete, update FIFO
    CANRXFIFO_OutDone();
    return TRUE; // message received
  }
  return FALSE; // no message received
}



/****************************************************************
DOES:     This function implements a CAN transmit FIFO. With each
          function call a message is added to the FIFO.
RETURNS: 1 Message was added to the transmit FIFO
         0 If FIFO is full, message was not added
****************************************************************/
UNSIGNED8 CCHW_PushMessage_Internal (
  CAN_MSG *pTransmitBuf // Data structure with message to be send
  )
{
CAN_MSG *pDst; // Destination pointer

  CCHW_DisableCANrx();
  // Get next message space available in FIFO
  pDst = CANTXFIFO_GetInPtr();
  if (pDst != 0)
  {
    // Copy message to transmit queue
    memcpy(pDst,pTransmitBuf,sizeof(CAN_MSG));
    // Copy completed
    CANTXFIFO_InDone();
    CCHW_EnableCANrx();
    return TRUE;
  }
  // Overrun occured
  // ADD OWN CODE TO HANDLE OVERRUN ERROR !!!
  CCHW_EnableCANrx();
  return FALSE;
}
```

```
/*******************************************************************
DOES:     This function handles CANcrypt CAN message transmissions.
          Includes CANcrypt secure message handling.
RETURNS: 1 Message was added to the transmit FIFO
         0 If FIFO is full, message was not added
*******************************************************************/
UNSIGNED8 CCHW_PushMessage (
  CAN_MSG *pTransmitBuf // Data structure with message to be send
  )
{
UNSIGNED8 ret_val = FALSE;

#ifdef Cc_USE_SECURE_MSG

UNSIGNED8 sec;
CAN_MSG preamble;

  CCHW_DisableCANrx();
  // check secure transmit for this message
  sec = Cc_Process_secMsg_Tx(&CcH,&preamble,pTransmitBuf);
  if (sec == 1)
  { // secure message, send preamble and message
    CCHW_PushMessage_Internal(&preamble);
    ret_val = CCHW_PushMessage_Internal(pTransmitBuf);
  }

  else if (sec == 2)
  { // Message known, but we are not secure
    // do not add message to FIFO
  }

  else
  { // Message not known to CANcrypt, transmit
    ret_val = CCHW_PushMessage_Internal(pTransmitBuf);
  }

  CCHW_EnableCANrx();

#else

  ret_val = CCHW_PushMessage_Internal(pTransmitBuf);

#endif

  return ret_val;
}
```

```
/******************************************************************
DOES:     Transmission of a CANcrypt high priority message.
NOTE:     Must be transmit immediately, no delay. Bypasses Tx FIFO.
          ONLY REQUIRED FOR DIRECT BIT CLAIMING MODE
RETURNS: Nothing
******************************************************************/
UNSIGNED8 CCHW_TransmitNow (
  CAN_MSG *pMsg
  )
{
UNSIGNED32 *pAddr;
UNSIGNED32 *pCandata;

  pAddr = (UNSIGNED32 *) &LPC_CAN1->TFI2;

  // Write DLC
  *pAddr = ((UNSIGNED32) (pMsg->LEN)) << 16;

  // Write CAN ID
  pAddr++;
  *pAddr = pMsg->ID & 0x000007FFL;

  // Write first 4 data bytes
  pCandata = (UNSIGNED32 *) &(pMsg->BUF[0]);
  pAddr++;
  *pAddr = *pCandata;

  // Write second 4 data bytes
  pCandata++;
  pAddr++;
  *pAddr = *pCandata;

  // Write transmission request
  pAddr = (UNSIGNED32 *) &LPC_CAN1->CMR;
  *pAddr = 0x50; // Transmission Request Buf 2, self receipt

  return TRUE;
}
```

```
/**********************************************************************
DOES: CAN transmit handler, check if there is something to transmit.
     If there is something in the transmit queue / FIFO, and if the
     transmit buffer is available, then copy next message from queue
     to the transmit buffer.
**********************************************************************/
void MCOHW_CheckTxQueue (
  void
  )
{
UNSIGNED32 *pCSR;              // pointer into SFR space
CAN_MSG MEM_BUF *pMsg;
UNSIGNED32 *pAddr;
UNSIGNED32 *pCandata;

  // Set SFR pointer
  pCSR = (UNSIGNED32 *) &LPC_CAN1->SR;

  // Get next message from FIFO
  pMsg = CANTXFIFO_GetOutPtr();

  if ((*pCSR & 0x00000004L) && (pMsg != 0))
  { // Transmit Channel 1 is available and message is in queue

    // Get register address to write to
    pAddr = (UNSIGNED32 *) &LPC_CAN1->TFI1;

    // Write DLC
    *pAddr = ((UNSIGNED32) (pMsg->LEN)) << 16;

    // Write CAN ID
    pAddr++;
    *pAddr = pMsg->ID & 0x000007FFL;

    // Write first 4 data bytes
    pCandata = (UNSIGNED32 *) &(pMsg->BUF[0]);
    pAddr++;
    *pAddr = *pCandata;
    // Write second 4 data bytes
    pCandata++;
    pAddr++;
    *pAddr = *pCandata;

    // maintain pseudo random
    *pPseudoRand += *pCandata;

    // Write transmission request
    pAddr = (UNSIGNED32 *) &LPC_CAN1->CMR ;
    *pAddr = 0x21; // Transmission Request Buf 1

    // Update Out pointer
    CANTXFIFO_OutDone();
  }
  // maintain pseudo random
  *pPseudoRand += gTimCnt;
}
```

CAN receive interrupt handler

```
/********************************************************************
DOES:     CAN receive interrupt handler
********************************************************************/
void MCOHW_CAN1ISR_Rx (
  void
  )
{
CAN_MSG MEM_BUF *pDst;
UNSIGNED32 *pDest;
UNSIGNED8 status;

  // maintain pseudo random
  *pPseudoRand += LPC_CAN1->RDA;

  if (!(LPC_CAN1->RFS & 0xC0000400L))
  { // 11-bit ID, no RTR, matched a filter
    // secCAN message handler, copy message into record
    // copy ID
    CANrx.ID = LPC_CAN1->RID & 0x000007FFL;
    // copy DLC
    CANrx.LEN = (LPC_CAN1->RFS & 0x000F0000L) >> 16;
    // copy data
    pDest = (UNSIGNED32 *) &(CANrx.BUF[0]);
    *pDest++ = LPC_CAN1->RDA;
    *pDest = LPC_CAN1->RDB;
    // copying is all done

    // check if this is a CANcrypt related message
    status = Cc_Process_Rx(&CcH,&CANrx);
    if (status == FALSE)
    { // not processed by CANcrypt
      // initialize destination pointer into FIFO
      pDst = CANRXFIFO_GetInPtr();
      if (pDst != 0)
      { // FIFO available
        memcpy(pDst,&CANrx,sizeof(CAN_MSG));
        // copying is all done
        // copying is all done
        CANRXFIFO_InDone();
      }
      else
      { // overrun, message lost
        // ADD OWN CODE TO HANDLE OVERRUN ERROR !!!
      }

    }
  }

  LPC_CAN1->CMR = 0x04; // release receive buffer

  return;
}
```

```
/*********************************************************************
DOES:    CAN error interrupt handler
*********************************************************************/
void MCOHW_CAN1ISR_Err (
  void
  )
{

  // INSERT APPLICATION SPECIFIC CODE AS NEEDED

  if (LPC_CAN1->GSR & 0x80)
  { // Bus off
    // Clear reset bit set by bus-off condition
    LPC_CAN1->MOD = 0;
  }
}



/*********************************************************************
DOES:    Main CAN interrupt handler, calls other CAN handlers
*********************************************************************/
void __irq CAN_IRQHandler (
  void
  )
{
        volatile UNSIGNED32 temp;

  if ( LPC_CAN1->GSR & (1 << 0 ) )
  {
    MCOHW_CAN1ISR_Rx();
  }
  if ( LPC_CAN1->GSR & (3 << 6 ) )
  {
    MCOHW_CAN1ISR_Err();
  }
  // Clear interrupt
  temp = LPC_CAN1->ICR;

  return;
}
```

Timer interrupt handler, timer and timeout

```
/*********************************************************************
DOES:    Timer interrupt handler (1ms)
*********************************************************************/
void __irq TIMER0_IRQHandler (
  void
  )
{
  LPC_TIM0->IR = 1; // Clear interrupt flag
  gTimCnt++; // increment global timer counter
  MCOHW_CheckTxQueue(); // check if something is in the Tx queue
}
```

```
/********************************************************************
DOES:    This function reads a 1 millisecond timer tick. The timer tick
         must be a UNSIGNED16 and must be incremented once per millisecond.
RETURNS: 1 millisecond timer tick
********************************************************************/
UNSIGNED16 CCHW_GetTime (void)
{
  return gTimCnt;
}


/********************************************************************
DOES:    This function compares a UNSIGNED16 timestamp to the internal
         timer tick and returns 1 if the timestamp expired/passed.
RETURNS: 1 if timestamp expired/passed
         0 if timestamp is not yet reached
NOTES:   The maximum timer runtime measurable is 0x8000 (about 32 seconds).
         For the usage in MicroCANopen that is sufficient.
********************************************************************/
UNSIGNED8 CCHW_IsTimeExpired (
  UNSIGNED16 timestamp
  )
{
UNSIGNED16 time_now;

  time_now = gTimCnt;
  if (time_now >= timestamp)
  {
    if ((time_now - timestamp) < 0x8000)
      return 1;
    else
      return 0;
  }
  else
  {
    if ((timestamp - time_now) >= 0x8000)
      return 1;
    else
      return 0;
  }
}
```

```
#ifdef CANcrypt_DEBUG
void DebugOutCAN(
  UNSIGNED16 id, UNSIGNED8 len,
  UNSIGNED8 d0, UNSIGNED8 d1, UNSIGNED8 d2, UNSIGNED8 d3,
  UNSIGNED8 d4, UNSIGNED8 d5, UNSIGNED8 d6, UNSIGNED8 d7
  )
{
CAN_MSG dbg;

  dbg.ID = Cc_CANID_DEBUG + id - 1;
  dbg.LEN = len;
  dbg.BUF[0] = d0;
  dbg.BUF[1] = d1;
  dbg.BUF[2] = d2;
  dbg.BUF[3] = d3;
  dbg.BUF[4] = d4;
  dbg.BUF[5] = d5;
  dbg.BUF[6] = d6;
  dbg.BUF[7] = d7;
  CCHW_PushMessage(&dbg);
}
#endif

/*---------------------- END OF FILE ----------------------------*/
```

8.5.3   NXP LPC17xx main – secure grouping example

The example file below shows a minimal CANcrypt setup, here for grouping. This example sends CANopen style bootup and heartbeat messages in the background.

The event call-back *Cccb_Event()* handles the grouping and secure heartbeat events. The *Cc_Process_Tick()* function is called in the background loop.

```
/*****************************************************************
MODULE:    main_xxxx.c, CANcrypt Demo
CONTAINS:  Main loop for CANcrypt demo
COPYRIGHT: Embedded Systems Academy GmbH, 2016-2017
HOME:      www.esacademy.com/cancrypt
LICENSE:   FOR EDUCATIONAL AND EVALUATION PURPOSE ONLY!
CONTACT:   Contact info@esacademy.de for other available licenses

Unless required by applicable law or agreed to in writing, software
distributed under the License is distributed on an "AS IS" BASIS,
WITHOUT WARRANTIES OR CONDITIONS OF ANY KIND, either express or implied.

VERSION:   0.10, 19-JAN-2017
*****************************************************************/

#include "CANcrypt_includes.h"
#include "LPC17xx.h" // LPC17xx Peripheral Registers

#ifdef Cc_USE_DIGOUT
#include "Cc_digout.h"
#endif

/*****************************************************************
GLOBAL VARIABLES
*****************************************************************/

// CANcrypt handle
Cc_HANDLE CcH;

// Counter for procss control
UNSIGNED32 counter = 0;

// secure message transmit repetiton time
UNSIGNED16 msg_time;

// Counter for secure heartbeats
UNSIGNED32 secHBcnt = 0;

#ifdef Cc_USE_SECURE_MSG
// secure transmit message table
Cc_SEC_MSG_TABLE_ENTRY TxSecMg[2] = {0x0180+NODEID, 1, 4, 0, NODEID,
                                     0xFFFFul, 0xFF, 0xFF, 0xFF, 0xFF};

UNSIGNED8 TxTrk[1];
```

```c
// secure receive message table
Cc_SEC_MSG_TABLE_ENTRY RxSecMg[4] = {
#if (NODEID != 2)
                                    0x0182, 1, 4, 0, 2,
#endif
#if (NODEID != 3)
                                    0x0183, 1, 4, 0, 3,
#endif
#if (NODEID != 7)
                                    0x0187, 1, 4, 0, 7,
#endif
                                    0xFFFFFul, 0xFF, 0xFF, 0xFF, 0xFF};
Cc_SEC_MSG_TRACK_ENTRY RxTrk[4];
#endif

/*********************************************************************
DOES:    Reports a CANcrypt system event to the application
RETURNS: Nothing
*********************************************************************/
void Cccb_Event(
  UNSIGNED8 event,              // Cc_EVENT_xxx
  UNSIGNED32 param1,            // see Cc_EVENT_xxx definitions
  UNSIGNED32 param2,
  UNSIGNED32 param3
)
{
CAN_MSG can;

  switch (event)
  {

    case Cc_EVENT_GROUP_INIT: // send CANopen HB, pre-operational
      can.ID = 0x700 + NODEID;
      can.BUF[0] = 0x7F;
      can.LEN = 1;
      CCHW_PushMessage(&can);
      break;

    case Cc_EVENT_GROUPED: // send CANopen HB, operational
      can.ID = 0x700 + NODEID;
      can.BUF[0] = 0x05;
      can.LEN = 1;
      CCHW_PushMessage(&can);
      break;
```

```
    case Cc_ERR_SECHB_FAIL: // send CANopen emergency
      can.ID = 0x80 + NODEID;
      can.BUF[0] = 0x33;
      can.BUF[1] = 0x33;
      can.BUF[2] = 0x01;
      can.BUF[3] = param1;
      can.BUF[4] = 0;
      can.BUF[5] = 0;
      can.BUF[6] = 0;
      can.BUF[7] = 0;
      can.LEN = 8;
      CCHW_PushMessage(&can);
      // start again
      counter = 0;
      break;

    case Cc_ERR_GROUP_LOST: // send CANopen emergency
      can.ID = 0x80 + NODEID;
      can.BUF[0] = 0x22;
      can.BUF[1] = 0x92;
      can.BUF[2] = 0x01;
      can.BUF[3] = param1;
      can.BUF[4] = param1 >> 8;
      can.BUF[5] = param2;
      can.BUF[6] = param2 >> 8;
      can.BUF[7] = 0;
      can.LEN = 8;
      CCHW_PushMessage(&can);
      // start again
      counter = 0;
      break;

    case Cc_EVENT_HBSECURED:
      can.ID = 0x700 + NODEID;
      can.BUF[0] = 0x05;
      can.LEN = 1;
      CCHW_PushMessage(&can);
      secHBcnt++;
      break;

    default:
      break;

  }
}


/*******************************************************************
DOES:    Trigger next transmit of a secure message
RETURNS: Nothing
*******************************************************************/
void TriggerSecureTx()
{
```

```
  static UNSIGNED8 cnt;
  CAN_MSG tx;

  cnt++;
  tx.BUF[0] = 0x33;
  tx.BUF[1] = NODEID;
  tx.BUF[2] = cnt;
  tx.BUF[3] = NODEID;
  tx.BUF[4] = 0xCC;
  tx.LEN = 5;
  // Demo only, transmit copy first
  tx.ID = 0x480 + NODEID;
  CCHW_PushMessage(&tx);
  // now transmit original
  tx.ID = 0x180 + NODEID;
  CCHW_PushMessage(&tx);
}


/*******************************************************************
DOES:    The main function
RETURNS: nothing
*******************************************************************/
int main(
  void
  )
{
CAN_MSG CANapp;

  SystemInit();

  // Init CAN interface
  CCHW_Init(125);

#ifdef Cc_USE_DIGOUT
  LPC_GPIO2->FIODIR = 0x0000007C; // P2 LED Outputs
  Cc_DIGOUT1_OFF;
  Cc_DIGOUT2_OFF;
#endif

  // initial repetion time for secure transmit
  msg_time = CCHW_GetTime() + 4000;

  // foreground loop
  while(1)
  {
    if (counter == 0)
    {
      // send boot up message
      CANapp.ID = 0x700 + NODEID;
      CANapp.BUF[0] = 0;
      CANapp.LEN = 1;
      CCHW_PushMessage(&CANapp);
    }
```

```
    // use counter as control as when to enable secCAN
    if (counter == 1000)
    {
      srand(CCHW_Rand()); // use seed with pseudo HW random
      Cc_Restart(&CcH,0,Cc_GROUP_CTRL_RESTART,Cccb_Event,
                 CCHW_PushMessage,CCHW_TransmitNow,NULL,NULL);
#ifdef Cc_USE_SECURE_MSG
      Cc_Load_Sec_Msg_Table(&CcH,RxSecMg,RxTrk,TxSecMg,TxTrk);
#endif
    }

    if (counter >= 1000)
    {
      Cc_Process_Tick(&CcH);
      // secure tranmit
      if (CCHW_IsTimeExpired(msg_time))
      {
        TriggerSecureTx();
        // next trigger ranges randomly from 100ms to about 350ms
        msg_time = CCHW_GetTime() + 100 + (CCHW_Rand() & 0xFF);
      }
    }

    // stay here
    if (counter < 10000)
    {
      counter++;
    }

    // Process incoming CAN messages for application
    if (CCHW_PullMessage(&CANapp))
    {
      if ((CANapp.ID == 0) && (CANapp.BUF[0] == 0x81))
      { // CANopen NMT Reset, here start over
        CANapp.ID = 0xFFFF;
        counter = 0;
        // initial repetion time for secure transmit
        msg_time = CCHW_GetTime() + 4000;
      }
      else if ((CANapp.ID & 0x180) == 0x180)
      { // secure PDO received, transmit echo
        CANapp.ID = 0x280 + NODEID;
        CCHW_PushMessage(&CANapp);
      }
    }

  } // end of while(1)
} // end of main
```

"The difference between the right word and the almost right word is the difference between lightning and a lightning bug."

"Martin Wak"

9 CANcrypt and CANopen

The default configuration of CANcrypt can be directly used with CANopen. The CAN message identifiers do not collide with the CANopen protocol. For more details on CANopen, see (Pfeiffer, 2003) and (CiA 301, V4.2 2007).

Alternatively, minimal CANcrypt functionality could also be implemented by packing the CANcrypt required data into existing CANopen messages. The following section shows how this information can be integrated into the CANopen emergency message.

In the following sections, we examine the security requirements for the different CANopen communication services.

Last follows a summary of al CANcrypt parameters that can be mapped to the CANopen Object Dictionary.

9.1   Grouping support with the emergency message

Applications that require minimal security can add grouping and authentication using only the CANopen emergency messages. No further CAN messages are required.

CANopen defines one emergency message for each node. Nodes may produce this message to indicate emergencies and end of emergencies. The emergency code is 16 bits. Values less than 100h indicate the end of an emergency.

The message can also include five manufacturer-specific information bytes. CANcrypt packs the data required by CANcrypt into these bytes.

The tables below show the emergency codes and their meanings. The first table lists emergency codes below 100h, indicating no error. The messages include acknowledgement, grouping progress, and the secure heartbeat.

EMCY code	Description
0090h	Generic CANcrypt acknowledge contents: status byte, key ID, key length
0091h	Grouping request, still nodes missing contents: status byte, key ID, key length, 16-bit random
0092h	Grouping request, no more nodes missing contents: status byte, key ID, key length, 16-bit random
0093h	Secure heartbeat, authenticates all previous messages contents: status byte, 24-bit random value, checksum

CANOPEN EMERGENCY CODES USED FOR CANCRYPT CONTROL

The second table lists the alerts with emergency codes greater than 100h, indicating an error or alert.

EMCY code	Description
8300h	Generic CANcrypt sequence abort contents: status byte, key ID, key length
8310h	CANcrypt alert event contents: status byte, 16-bit CANcrypt error code
8320h	Generic ungroup request, communication no longer secure contents: status byte
8330h	Intruder detected alert, detected by this device contents: status byte

CANOPEN EMERGENCY CODES FOR CANCRYPT ALERTS

9.2 The NMT Master Message

CANopen's NMT Master message is very powerful. It contains commands for individual devices or all devices to change operating modes (stop operation) and commands to initiate a device reset. Without authentication, the injection of just this single message can cause major harm to a system. Once security is added to a CANopen system, the NMT Master message is one of the first messages requiring authentication.

In any CANopen CANcrypt system, the NMT Master participates in the grouping process and supports the generation of the CANcrypt secure heartbeat.

Before processing a received NMT Master message, a device must delay processing the message until receiving the next secure heartbeat from the NMT Master. Only reception and verification of the secure heartbeat authenticates all messages sent by the NMT master since its last secure heartbeat.

Note: due to delays in transmit and receive FIFO handling, for 100% security, receiver would need to wait for the second secure heartbeat.

On receiving a secure heartbeat timeout or receiving a CANcrypt alert from the CANopen NMT Master, a device shall not execute the last received NMT command as it was not authenticated.

9.3   Securing PDOs

Process Data Objects (PDOs) are CAN messages with process data that provide a multicast mechanism to send a message to multiple consumers. These messages are best authenticated using the secure heartbeat. All nodes transmitting PDOs that require authentication and all nodes receiving PDOs must participate in secure heartbeat generation and consumption.

A received PDO may be regarded as authenticated on receiving the following secure heartbeat from the node that transmitted the PDO.

Encryption of PDOs is currently only supported for paired CANcrypt devices. In this case, only the two paired devices can produce and consume encrypted PDOs. All other devices do not have access to the encrypted data.

9.4   Securing SDOs

Service Data Objects (SDOs) are CAN messages with configuration or service data that provide point-to-point communication, typically between the CANopen Manager and the devices. These messages can also be authenticated using the secure heartbeat.

A communication mode like SDO is already provided by CANcrypt. The generic data access protocol provides the option to read or write single object dictionary entries of up to 4 bytes in length. However, this access method is limited to the configurator.

A more generic access with both authentication and encryption support can be achieved by pairing the two devices that are communicating and adding the SDO messages to the list of secure messages. This method supports all SDO communication methods, including segmented or blocked transfers.

9.5 Object Dictionary entries for CANcrypt

The object dictionary entries required for the configuration of CANcrypt are divided into three sections: the extended identification and status information, the key management and configuration of the secure messages.

9.5.1 Extended identification and status

These entries are all read-only. The entries at 5EF0h are duplicates from what is available through other CANcrypt protocols. The intention here is to provide this data in a CANopen compatible way. If CANopen is used on this device, then this provides an alternate access to this information.

Use	Index	Subindex	Type	Access
Vendor ID (0: no CANopen Vendor)	1018h	1	UNSIGNED32	RO
Product code	1018h	2	UNSIGNED32	RO
Revision number	1018h	3	UNSIGNED32	RO
Serial number (0 if not used)	1018h	4	UNSIGNED32	RO
CANcrypt version and support	5EF0h	1	UNSIGNED16	RO
CANcrypt address	5EF0h	2	UNSIGNED8	RO
CANcrypt status	5EF0h	3	UNSIGNED8	RO
Current key ID and length	5EF0h	4	UNSIGNED16	RO

CANOPEN OBJECT DICTIONARY ENTRIES FOR IDENTIFICATION AND CANCRYPT STATUS

9.5.2 Key management

There is no direct access to the keys. They cannot be read using CAN communication. Using the CANcrypt generic access protocol, a configurator can instruct a device to erase a key and to save the current dynamic key as a permanent key.

Name	Index	Subindex	Type	Access	M/O
User key access	5EF1h	1	UNSIGNED8	R/W	O
Identifier for above	5EF2h	1	UNSIGNED32	R/W	O
Owner key access	5EF1h	2	UNSIGNED8	R/W	O
Identifier for above	5EF2h	2	UNSIGNED32	R/W	O
System integrator key access	5EF1h	3	UNSIGNED8	R/W	O
Identifier for above	5EF2h	3	UNSIGNED32	R/W	O
Manufacturer key access	5EF1h	4	UNSIGNED8	R/W	O
Identifier for above	5EF2h	4	UNSIGNED32	R/W	O

CANOPEN OBJECT DICTIONARY ENTRIES FOR KEY MANAGEMENT

The 32-bit key identifier at 5EF2h may be written by the configurator when saving a key. This allows a configurator to later verify which key was saved here.

Reading the access field at 5EF1h provides the following information:

> 11h: no key stored at this location

> 33h: a key is stored at this location (see 5EF2h for identifier)

> FFh: this entry was never used before

On writing the access fields, the following values are accepted:

> DDh: delete the key stored at this location

> 5Dh: save the current dynamic key at this location

9.5.3 Secure messages

For CANcrypt systems supporting secure messaging, tables are required with one entry for each CAN message to be handled securely. There is one table for transmit (secure messages transmitted) and one table for receive.

Name	Index	Subindex	Type	Access
Secure Transmit Message List CAN ID	5EE3h	1-255	UNSIGNED32	RO, R/W
Secure Transmit Message List Configuration (first encrypted, number of, functionality, method, producer)	5EE5h	1-255	UNSIGNED32	RO, R/W

CANOPEN OBJECT DICTIONARY ENTRIES FOR
SECURE MESSAGE TRANSMIT LIST

Name	Index	Subindex	Type	Access
Secure Receive Message List CAN ID	5EE6h	1-255	UNSIGNED32	RO, R/W
Secure Receive Message List Configuration (first encrypted, number of, functionality, method, producer)	5EE8h	1-255	UNSIGNED32	RO, R/W

CANOPEN OBJECT DICTIONARY ENTRIES FOR
SECURE MESSAGE RECEIVE LIST

If the CANcrypt device is a CANopen device, then carefully consider if these entries are made visible on the CANopen side as that would potentially allow unsecured access to the table.

We recommend that if the tables are made visible with regular CANopen accesses, then they should be read-only. Writing the table should only be allowed for a paired configurator using the secured generic write access.

10   CANcrypt attack vectors

In this chapter we examine the remaining attack vectors of CANcrypt.

10.1   Randomness

Wherever random numbers are used, a typical attack vector is to assume that the numbers are not random and to determine a pattern. The random numbers used by the paired devices must be "reasonably good", and they must differ with every power cycle.

This might be a challenge for many microcontroller systems. The random functions provided by C compilers are typically not good enough to ensure "randomness". If these are used, it is important to update the random seed used by this function as often as possible.

One possible option is to maintain and update a seed value that is constantly updated (using any arithmetic function) depending on inputs from A/D converters (especially highest resolution bits), high-resolution timers (measuring timestamps of CAN messages received).

Latest microcontrollers more and more often have a dedicated random number generator. Where available, these should be used.

10.2   Creation and storing of permanent keys

The phase where keys are stored permanently is crucial. Even if an intruder cannot see keys generated by the CANcrypt methods, a potential intruder present at this stage would be a security concern. So whenever initial pairing keys (no matter if by the manufacturer or system integrator) are generated and stored, extra precautions should be taken to ensure that no intruder is present.

The precautions can include verifying that only those devices that are physically connected to CAN are required for this process.

Some microcontrollers offer dedicated memory locations for storage of security information. Where available, this area should be used to store the key(s) with the highest priority.

If "regular" FLASH memory is used, ensure that the FLASH protection bits are set. Specific functionality of these depends on the manufacturer. Usually these can be configured in a way that only the code executed from the FLASH memory can read-access this memory. Where possible, protect the FLASH from being read from RAM and FLASH programming ports/interfaces.

10.3 Debug and bootloader interfaces

Microcontrollers with an enabled JTAG or SWD interface for debugging or an internal bootloader to reprogram the device can be very vulnerable to attacks. This is especially true when these accesses are also made available through other communication interfaces such as UART, USB, or even CAN. Using an appropriate debugger or programming tool, an intruder can manipulate memory contents and inject code.

For highest security, all such interfaces should be internally disabled. Bootloaders should only be enabled if they offer their own security levels such as protection from unauthorized activation and supporting flashing only of authorized files.

10.4 Read/write access to CAN system

This section summarizes attacks where the intruder has full CAN level access and can receive all messages and inject messages at will. The access could be through a physically connected sniffer or a remote access device.

In CAN networks, a typical attack involves recording messages and replaying them. If the messages exchanged after power up are always the same, an attacker could fake initial messages by replaying them. However, due to the dynamic, random key changes, a hacker would probably look at alternate methods first. Examples include trying to activate a boot loader or re-flashing a device with new code.

10.5 Ability to physically remove/replace devices

With physical removal, an attacker has the ability to reprogram (re-flash) a connected device.

If one of the paired or grouped devices is removed and replaced with a tampered device, the tampered device can participate in the key generation or grouping cycles. However, all pairing and grouping functions also use elements of a permanent key. With endless time and re-tries an attacker might be able to determine parts of the permanent key. To make this attack vector less attractive, CANcrypt uses incremental delays on pairing/grouping re-tries.

10.6 Signal level access to CAN and PCBs

The result that can be achieved with signal-level access heavily depends on many factors. With PCB access to a CAN transceiver, an attacker would be able to see the bit generation. If session pairing is used, the attacker can see all keys generated. If permanent pairing is used, an attacker would still be able to learn changes to the dynamic key over time and eventually have a copy of the dynamic key. Although this level of attack is theoretically possible, any active attack trying to manipulate or fake data would still be difficult as it would require hard, real-time reaction.

Note that this attack option is not available if the paired devices use controllers with integrated transceivers such as some of the NXP LPC11Cxx devices.

10.7 Computational Cryptanalysis

There are various brute force computational methods available that can also be applied to the CANcrypt signatures or secure messages. The success rate of such methods also depends on data sizes available for the analysis. Here CANcrypt is challenging due to the use of dynamic keys and one-time pads. And on top of it customizable security functions. So an attacker cannot really be sure what he is dealing with.

10.8 Summary

At this point we are not aware of any "promising" attack vectors on the CAN/CANopen level in a CANCrypt system. That includes remote access (for example through a hacked gateway) as well as direct access with a CAN sniffer utility.

Unlimited physical access to the network and the devices (PCB level) may produce a result over time (being able to listen to communication over time). However, actively participating in the communication will be very difficult and only of limited use as paired devices will recognize the additional communication and disconnect.

At Embedded Systems Academy we are currently evaluating the possibility to set up a bounty program. Visit our web page for latest information.

It will be interesting to see which cryptanalysis methods produce what kind of results in which time frame.

Appendix A - Provided Demos and Files

You can find the latest CANcrypt demos at www.esacademy.com/cancrypt. The initial release consists of 5 demo projects.

CANcrypt API

This project does not include any real code, it is just a collection of all definition files for types and functions.

CANcrypt Monitoring Demo

This demo monitors the CAN transmit and receive ports. It collects a list of all CAN message IDs transmitted. If then a message is received that is in the list, an unexpected message alert is generated as we must assume that this message was injected by someone else.

The directory "trace" includes a CAN trace recording "Unexpected_N7.xlsx" of a single CANopen device with the node ID 7 using this functionality. Injected messages cause an alert.

CANcrypt Grouping Demo

This demo uses multiple devices depending on configuration. Which devices are expected in the group is determined by the Ccnvol_GetGroupInfo() call-back function implemented in the mcohw_xxxx.c module.

This demo was tested with up to four devices (2, 3, 7 and 15).

On power-up devices try to group with each other. Once grouped, the devices start the transmission of a cyclic secure heartbeat.

The directory "trace" includes a CAN trace recording "Group_N2_N3_N7.xlsx" of a boot-up of a system with CANcrypt devices 2, 3 and 7. First all three send a grouping request followed by a grouping confirmation. Once grouped (see status byte is 08h), the start producing the secure heartbeat.

Comparison of code sizes and execution times

Code sizes		LPC2368 ARM7 (ARM32)	LPC11C24 Cortex M0 (Thumb-2)	LPC1768 Cortex M3 (Thumb-2)
Grouping example - BASIC, REGULAR		3248 Bytes	2002 Bytes	2116 Bytes

The code sizes listed above are for the three modules Cc_system.c, Cc_Grouping.c and Cc_user_default.c. Code was compiled with Keil/ARM C compiler version V5.06, optimization level 3 – without cross module optimization.

Execution times in microseconds, 128 bit permanent key, 64 bit dynamic key	LPC2368 ARM7 (48 Mhz)	LPC11C24 Cortex M0 (48 Mhz)	LPC1768 Cortex M3 (48 Mhz)	LPC1768 Cortex M3 (72 Mhz)
Ccuser_ExpandRandom()	1.4	2.2	1.4	1.0
Ccuser_Mix64() - BASIC	7.3	9.2	6.4	4.2
Ccuser_Mix64() - REGULAR	25.2	31.3	20.1	13.4
One-time pad generation - BASIC	9.7	12.5	8.7	5.8
One-time pad generation - REGULAR	27.6	34.7	22.5	15.0
Generate / Verify Signature	0.5	0.7	0.4	0.3

The code execution times above are in microseconds. Preferably a new one-time pad is generated with every secure message. Generating a new secure heartbeat also requires generating a one-time pad (calls Ccuser_ExpandRandom() and Ccuser_Mix64(). In comparison, the time required for the actual generation and comparison of the signature are minimal.

Execution times in microseconds, 256 bit permanent key, 128 bit dynamic key	LPC2368 ARM7 (48 Mhz)	LPC11C24 Cortex M0 (48 Mhz)	LPC1768 Cortex M3 (48 Mhz)	LPC1768 Cortex M3 (72 Mhz)
Ccuser_ExpandRandom()	2.7	3.8	2.5	1.7
Ccuser_Mix64() - BASIC	18.7	22.9	15.9	10.6
Ccuser_Mix64() - REGULAR	52.8	66.2	43.5	29.0
One-time pad generation - BASIC	22.5	27.9	19.5	13.0
One-time pad generation - REGULAR	65.7	71.3	47.0	31.3
Generate / Verify Signature	0.5	0.7	0.4	0.3

This second table shows the timings for longer keys used. Here 256-bit permanent key and 128-bit dynamic key.

CANcrypt Key Generation Demo

This demo uses 2 devices: the CANcrypt configurator (device 1) and a CANcrypt device (here device 3). After power-up both devices send a CANopen compatible bootup message. Then the configurator starts a single key generation cycle. The bit claiming method is used, so that by just watching the CAN traffic one cannot determine the key generated.

The directory "trace" includes several CAN trace recordings for generation of a 32-bit key using different parameters.

KeyGen32_medium_delay_2claim.xlsx
Uses medium timing, bit-claiming messages are transmitted with a random delay and two bit claiming messages.

KeyGen32_medium_delay_16claim.xlsx
Uses medium timing, bit-claiming messages are transmitted with a random delay and 16 bit claiming messages.

KeyGen32_fast_delay_2claim.xlsx
Uses fast timing, bit-claiming messages are transmitted with a random delay and two bit claiming messages.

KeyGen32_fast_delay_16claim.xlsx
Uses fast timing, bit-claiming messages are transmitted with a random delay and 16 bit claiming messages.

KeyGen32_fast_direct_16claim.xlsx
Uses fast timing, bit-claiming messages are transmitted directly after the trigger with no delay and 16 bit-claiming messages.

CANcrypt Pairing and Secure Messaging Demo

This demo uses 2 devices: the CANcrypt configurator (device 1) and a CANcrypt device (here device 3). After power-up both devices send a CANopen compatible bootup message. Then the configurator starts the pairing process.

Once paired, the configurator cyclically runs bit claiming cycles to update the shared dynamic key.

About two seconds after pairing the devices start exchanging secure CAN messages (181h and 183h) that are encrypted and authenticated. Received secure messages are echoed using a different CAN ID (281h and 283h) so that their clear readable contents can also been seen on the CAN bus.

The directory "trace" includes a CAN trace recording "PairedAndSecure.xlsx" of a boot-up of a system with CANcrypt device 3 and the configurator (device 1). The configurator sends a pairing request and device 3 sends the pairing response, then both transmit the pairing confirmation. The configurator now cyclically starts a new bit-generation cycle to generate a single shared secret bit that both device use to update their dynamic shared key.

After about two seconds, both devices start transmitting secure messages, 181h and 183h. Each occurrence is a pair, first the preamble, followed by the secured data. In this demo configuration, the first two bytes of the data remain unencrypted. Data length is 5, the unused bytes are filled with random data.

Once a secure message is received, the receiving device re-transmits it unencrypted as an "echo" (281h and 283h) so that we can see the original data in the trace.

CANcrypt Grouping and Secure Messaging Demo

This demo uses three devices: 2, 3 and 7. After power-up the devices group and start generating the secure heartbeat which also updates the shared dynamic key.

A few seconds into operation, the devices each start transmitting a secured message which is received by the other two devices.

For better visualization, each message is first transmitted in plain text using the CAN ID 480h plus the device address. Then it is transmitted again securely, this time using the CAN ID 180h plus device address. On the CAN bus we then see two messages, first the pre-amble, then the data. In the configuration used, the first byte is NOT encrypted, only the following four bytes are.

The secure messages are received by the other two devices. It gets decrypted and if authenticated, the message is re-transmitted using the CAN ID 280h plus device address. So for each secure message, two confirmations are generated.

"The single biggest problem in communication is the illusion that it has taken place."

"Roger Shrew Bandage"

Appendix B - Selected Sources

This appendix contains selected C source listings from the demo code published at www.esacademy.com/cancrypt. The files shown here contain the most important declarations and definitions as well as the default implementation of the customizable security functions.

User types: Cc_user_types.h

This file contains the main types used by CANcrypt. The notation for the standard variable types is taken from the CANopen standard. Note that the data type used for the CAN message ID can be selected by CAN_ID_SIZE, set this to 16, if in your CAN system only 11-bit CAN identifiers are used. Set it to 32 if you also use the extended 29-bit CAN message identifiers in your system.

```
/***************************************************************
MODULE:    Cc_user_types.h, target specific user types
CONTAINS:  The user specific global type definitions for CANcrypt
AUTHOR:    2017 Embedded Systems Academy, GmbH
HOME:      www.esacademy.com/cancrypt

  Licensed under the Apache License, Version 2.0 (the "License");
  you may not use this file except in compliance with the License.
  You may obtain a copy of the License at
  www.apache.org/licenses/LICENSE-2.0

  Unless required by applicable law or agreed to in writing, software
  distributed under the License is distributed on an "AS IS" BASIS,
  WITHOUT WARRANTIES OR CONDITIONS OF ANY KIND, either express or
  implied. See the License for the specific language governing
  permissions and limitations under the License.

VERSION:   0.10, 19-JAN-2017
***************************************************************/

#ifndef _CC_USER_TYPES_H
#define _CC_USER_TYPES_H


/***************************************************************
DEFINES: MEMORY TYPE OPTIMIZATION
***************************************************************/
#define MEM_CONST        const
#define MEM_FAR
#define MEM_BUF
```

```
/*********************************************************************
DEFINES: TRUE AND FALSE
*********************************************************************/
#ifndef TRUE
 #define TRUE          (1==1)
#endif
#ifndef FALSE
 #define FALSE         (!TRUE)
#endif


/*********************************************************************
TYPEDEF: CANOPEN DATA TYPES
*********************************************************************/
typedef unsigned char    UNSIGNED8;
typedef unsigned short   UNSIGNED16;
typedef unsigned int     UNSIGNED32;
typedef char             INTEGER8;
typedef short            INTEGER16;
typedef int              INTEGER32;


/*********************************************************************
TYPEDEF: CAN IDENTIFIER TYPE and size
*********************************************************************/
#define CAN_ID_SIZE 16

#if (CAN_ID_SIZE == 16)

typedef UNSIGNED16 COBID_TYPE;
#define COBID_DISABLED  0x8000U
#define COBID_RTR       0x4000U
#define COBID_EXT       0x2000U
#define COBID_MASKCTRL  0xE000U
#define COBID_MASKID    0x07FFU

#elif (CAN_ID_SIZE == 32)

typedef UNSIGNED32 COBID_TYPE;
#define COBID_DISABLED  0x80000000UL
#define COBID_RTR       0x40000000UL
#define COBID_EXT       0x20000000UL
#define COBID_MASKCTRL  0xE0000000UL
#define COBID_MASKID    0x1FFFFFFFUL

#else

#error "Only CAN_ID_SIZE 16 or 32 is possible"

#endif


#endif
/*---------------------- END OF FILE ----------------------------*/
```

Customizable user functions: Cc_user_default.c

The listing below shows the default Version 0.10 implementation of the customizable security functions. When not using the default, please set the security functionality to "custom". You may then edit and modify these functions depending on your requirements.

```
/*****************************************************************
MODULE:     Cc_user_default.c, CANcrypt Customizable Functions Demo
CONTAINS:   The default set of customizable functions for CANcrypt
COPYRIGHT:  Embedded Systems Academy GmbH, 2016-2017
HOME:       www.esacademy.com/cancrypt
LICENSE:    FOR EDUCATIONAL AND EVALUATION PURPOSE ONLY!
CONTACT:    Contact info@esacademy.de for other available licenses

Unless required by applicable law or agreed to in writing,
software distributed under the License is distributed on an
"AS IS" BASIS, WITHOUT WARRANTIES OR CONDITIONS OF ANY KIND,
either express or implied.

VERSION:    0.10, 19-JAN-2017
*****************************************************************/

#include "CANcrypt_includes.h"
#include <string.h>


/*****************************************************************
This file implements the CANcrypt customizable security functions for
the security modes basic, regular and advanced. The advanced mode
uses known, exisiting security algorithms not further explained.
*****************************************************************/
#if (Cc_FUNCTIONALITY >= Cc_SECFCT_CUSTOM)
 #error "Cc_FUNCTIONALITY must be basic, regular or advanced!"
#endif
#if (Cc_VERSION_NR != 0)
 #error "File implements version 0 of CANcrypt default functions"
#endif
#if (Cc_REVISION_NR != 1)
 #error "File implements revision 1 of CANcrypt default functions"
#endif
#if ((Cc_PERMKEY_LEN_BITS != 128) && (Cc_PERMKEY_LEN_BITS != 256))
 #error "This module is for a permanent key length of 128/256 bit"
#endif
#if ((Cc_KEY_LEN_BITS != 64) && (Cc_KEY_LEN_BITS != 128))
 #error "This module is for a dynamic key length of 64/128 bit"
#endif
#if (Cc_KEY_LEN_BITS != (Cc_PERMKEY_LEN_BITS / 2))
 #error "Permanent key must be double length of dynamic key"
#endif
```

```
/************************************************************************
Macros to rotate 32bit value right or left and a single mix up round
in add-rotate-xor (ARX) style as used by Speck cipher
************************************************************************/
#define ROR32(x,r) ( (x >> (r & 0x1F)) | (x << (32 - (r & 0x1F))) )
#define ROL32(x,l) ( (x << (l & 0x1F)) | (x >> (32 - (l & 0x1F))) )
#define MIXROUND32(a,b,k) (a=ROR32(a,8),a+=b,a^=b,b=ROL32(b,3),b^=a)


/************************************************************************
BOOK:     Section 6.1 "Collect random numbers"
DOES:     This function expands an array with a limited number of
          random bytes to an array of random bytes with the length
          of the current dynamic key.
RETURNS: nothing
************************************************************************/
void Ccuser_ExpandRandom(
  UNSIGNED32 *pkey,      // key input, length Cc_KEY_LEN8
  UNSIGNED32 *pdest,     // destination: array with length of dyn. key
  UNSIGNED32 *psrc       // array with zeros and random numbers (3*15)
)
{
UNSIGNED32 tmp = 0;
UNSIGNED8 lp_s = 0;
UNSIGNED8 lp_d = 0;
UNSIGNED8 all_ins_used = 0;
UNSIGNED8 all_outs_used = 0;

#ifdef Cc_USE_DIGOUT
  // performance measurement
  Cc_DIGOUT1_ON;
#endif

  // init destination with current key
  memcpy(pdest,pkey,Cc_KEY_LEN8);
  // init work variable with end of key
  tmp = pkey[Cc_KEY_LEN32-1];

  while ( (all_ins_used == 0) && (all_outs_used == 0) )
  { // repeat this until all values have been used
    // add random values to key
    tmp += psrc[lp_s];
    pdest[lp_d] += tmp;
    lp_d++;
    if (lp_d >= Cc_KEY_LEN32)
    { // end of destination reached
      lp_d = 0;
      all_outs_used = 1;
    }
    lp_s++;
    if (lp_s >= 12)
    { // end of source reached
      lp_s = 0;
      all_ins_used = 1;
    }
  }
```

```
#ifdef Cc_USE_DIGOUT
  // performance measurement
  Cc_DIGOUT1_OFF;
#endif

}


/*********************************************************************
BOOK:    Section 6.2 "Bit mixup"
DOES:    This function mixes the bits in a 64bit value by applying
         a Speck cipher. Used by key initialization functions and
         one-time pad generation.
MOTE:    Recommended number of rounds is 27
RETURNS: Value pmixed[] returns the mixed bits
*********************************************************************/
void Ccuser_Mix64(
  UNSIGNED32 pkey[2],     // key input, length Cc_PERMKEY_LEN_BITS
  UNSIGNED32 pdat[2],     // data input of 64 bit
  UNSIGNED32 pmixed[2],   // mixed bits output of 64 bit
  UNSIGNED8 rounds        // number of mixing rounds to execute
)
{
  UNSIGNED32 d0 = pdat[0];
  UNSIGNED32 d1 = pdat[1];
  UNSIGNED32 key[Cc_PERMKEY_LEN32];
  UNSIGNED8 lp;

  memcpy((UNSIGNED8 *) key,(UNSIGNED8 *) pkey,Cc_PERMKEY_LEN8);

  MIXROUND32(d1, d0, key[0]); // apply key
  for (lp = 0; lp < rounds - 1; lp++)
  { // execute a round
    // key expansion
    MIXROUND32(key[lp & (Cc_PERMKEY_LEN32-1)],key[0],lp);
    MIXROUND32(d1,d0,key[0]); // apply key
  }
  // return result
  pmixed[0] = d0;
  pmixed[1] = d1;
}


/*********************************************************************
BOOK:    Section 6.3 "Generate keys"
DOES:    Takes input from 2 keys and 1 factor to create a new key.
         Used to create a dynamic key from a permanent key using
         random input and a serial number.
         Used to create a one-time pad from a permanent and
         dynamic key and a counter.
RETURNS: TRUE if key initialization completed,
         FALSE if not possible due to parameters
*********************************************************************/
```

```c
UNSIGNED8 Ccuser_MakeKey(
  UNSIGNED32 *pin1,      // input 1: pointer to permanent key used
  UNSIGNED32 *pin2,      // input 2: pointer to 2nd input array
  UNSIGNED32 factor,     // input 3: optional, set zero if not used
                         // used for serial number, counter
  UNSIGNED32 *pout       // output: the dynamic key or one time pad
)
{
UNSIGNED8 lp;
UNSIGNED8 ret_val = FALSE;

#ifdef Cc_USE_DIGOUT
  // performance measurement
  Cc_DIGOUT1_ON;
#endif

  if ( (pin1 != NULL) && (pin2 != NULL) && (pout != NULL))
  { // parameter check ok

    // for length of key
    for (lp = 0; lp < Cc_KEY_LEN32; lp += 2)
    { // for length of keys
      // introduce factor
      pin2[lp] += factor;
      pin2[lp+1] ^= factor;
      // number of rounds depending on security level
      // Speck defined value for 64/128bit keys: 27
#if (Cc_FUNCTIONALITY == Cc_SECFCT_BASIC)
      Ccuser_Mix64(pin1,&(pin2[lp]),&(pout[lp]),7);
#elif (Cc_FUNCTIONALITY == Cc_SECFCT_REGULAR)
      Ccuser_Mix64(&(pin1[lp]),&(pin2[lp]),&(pout[lp]),27);
#elif (Cc_FUNCTIONALITY == Cc_SECFCT_ADVANCED)
      // Using AES, see below;
#else
  #error "Functionality selection not supported"
#endif
    }

#if (Cc_FUNCTIONALITY == Cc_SECFCT_ADVANCED)
  #if (Cc_KEY_LEN32 != 4)
    #error "key length not supported"
  #endif
  // in advanced mode, now use AES
  AESxxx_Encryption(pin1,pin2,pout);
#endif

    ret_val = TRUE;
  }

#ifdef Cc_USE_DIGOUT
  // performance measurement
  Cc_DIGOUT1_OFF;
#endif

  return ret_val;
}
```

```
/*********************************************************************
BOOK:    Section 6.4.1 "Pairing: Key update using a single bit"
DOES:    Called to update a dynamic key by introducing a new bit.
RETURNS: TRUE if key update completed,
         FALSE if not possible due to parameters
*********************************************************************/
UNSIGNED8 Ccuser_UpdateDynKey(
  UNSIGNED8 bit,        // new bit to introduce to dynamic key
  UNSIGNED32 *ppermkey, // pointer to permanent key used
  UNSIGNED32 *pdynkey   // pointer to dynamic key
)
{
UNSIGNED8 lp;
UNSIGNED8 *pB;
UNSIGNED8 ret_val = FALSE;

  if ((ppermkey != NULL) && (pdynkey != 0))
  {
    // save last key
    memcpy(&(pdynkey[Cc_KEY_LEN32]),pdynkey,Cc_KEY_LEN8);

    // shift entire dynamic key right by 1 bit
    for(lp=Cc_KEY_LEN32-1;lp>0;lp--)
    { // loop backwards, from highest to lowest
      // shift right
      pdynkey[lp] >>= 1;
      // merge in bit from next record
      pdynkey[lp] |= (pdynkey[lp-1] << 31);
    }
    pdynkey[0] >>= 1;

    // Check if bit needs flipping
    pB = (UNSIGNED8 *) (ppermkey);
    if ((pB[pdynkey[0] & (Cc_PERMKEY_LEN8-1)]) & 1)
    { // flip bit
      bit ^= 1;
#ifdef CANcrypt_DEBUG
      pB = (UNSIGNED8 *) (pdynkey);
      DebugOutCAN(NODEID,5,pB[0],pB[1],pB[2],pB[3],0x80+bit,0,0,0);
#endif
    }
#ifdef CANcrypt_DEBUG
    else
    {
      pB = (UNSIGNED8 *) (pdynkey);
      DebugOutCAN(NODEID,5,pB[0],pB[1],pB[2],pB[3],bit,0,0,0);
    }
#endif

    // add highest bit to new 32bit
    // ensure bit stuffing ???
    if (bit)
    { // set bit
      pdynkey[0] |= 0x80000000ul;
    }
```

```
    ret_val = TRUE;
  }

  return ret_val;
}


/*********************************************************************
BOOK:    Section 6.5.1 "Generate signature value"
DOES:    Generates a signature for this device
RETURNS: The 32bit signature
*********************************************************************/
UNSIGNED32 Ccuser_MakeSignature(
  UNSIGNED8 address,      // device ID (1-15) of this device
  UNSIGNED32 *pdyn,       // pointer to dynamic key used, uses 32bits
  UNSIGNED32 rnd          // new random value to introduce)
)
{
UNSIGNED32 secHB = 0;
UNSIGNED8 sum;
UNSIGNED8 *p8;

  if (pdyn != NULL)
  { // parameter check ok

    // initialize checksum with value from dynamic key
    p8 = (UNSIGNED8 *) pdyn;
    sum = p8[address&(Cc_KEY_LEN8-1)];

    // add first 3 bytes of checksum
    sum += (rnd & 0xFF);
    sum += ((rnd >> 8)& 0xFF);
    sum += ((rnd >> 16)& 0xFF);

    // prepare result
    secHB = sum;
    secHB <<= 24;
    secHB += (rnd & 0x00FFFFFFul);

    // encrypt with XOR
    secHB ^= *pdyn;

  }

  return secHB;
}
```

```
/*****************************************************************
BOOK:    Section 6.5.2 "Verify signature value"
DOES:    Verifies a signature received from a device
RETURNS: TRUE, if signature was verified
*****************************************************************/
UNSIGNED8 Ccuser_VerifySignature(
  UNSIGNED8 address,     // device ID (1-15) of device sending the
  UNSIGNED32 *sHB,       // signature
                         // on return, decrypted value is at location
  UNSIGNED32 *pdyn       // pointer to dynamic key used
)
{
UNSIGNED8 add;
UNSIGNED8 *p8;
UNSIGNED8 ret_val = FALSE;

#ifdef Cc_USE_DIGOUT
  // performance measurement
  Cc_DIGOUT2_ON;
#endif

  if (pdyn != NULL)
  { // parameter check ok
    // initialize checksum with value from dynamic key
    p8 = (UNSIGNED8 *) pdyn;
    add = p8[address&(Cc_KEY_LEN8-1)];

    // decrypt with XOR
    *sHB ^= *pdyn;

    // add first 3 bytes of checksum
    add += (*sHB & 0xFF);
    add += ((*sHB >> 8)& 0xFF);
    add += ((*sHB >> 16)& 0xFF);

    // verify result
    if (add == (*sHB >> 24))
    { // sum matches
      ret_val = TRUE;
    }

  }

#ifdef Cc_USE_DIGOUT
  // performance measurement
  Cc_DIGOUT2_OFF;
#endif

  return ret_val;
}
```

```
/*******************************************************************
BOOK:      Section 6.6.1 "Checksum init"
DOES:      Generates an initial value for the checksum, depending on
           a key passed.
RETURNS: 32 bit initial value
*******************************************************************/
UNSIGNED32 Ccuser_ChecksumInit(
  UNSIGNED32 *pkey        // pointer to a key (pad, dyn or perm)
)
{
UNSIGNED32 sum;

  if (pkey != NULL)
  {
    sum = pkey[0] ^ pkey[1];
  }
  else
  { // not good, consider reporting an error
    sum = 0x5BADC0DE;
  }

  return sum;
}


/*******************************************************************
BOOK:      Section 6.6.2 "Checksum step"
DOES:      Calculates a 16bit checksum, adding one value at the time
RETURNS: Checksum value in lowest 16bit, highest 16bit is carry-over
*******************************************************************/
UNSIGNED32 Ccuser_ChecksumStep16(
  UNSIGNED32 last,        // initial value or last calculated value
                          // higher 16bit may include a carry-over
  UNSIGNED16 *pdat        // next 16bit value to add
)
{
UNSIGNED32 sum;

  if (pdat != NULL)
  { // parameter check ok
    // Basic version: Fletcher style checksum
    sum = last + *pdat;
  }

  return sum;
}
```

```
/*********************************************************************
BOOK:    Section 6.6.3 "Checksum final"
DOES:    When checksum calculation is completed, merges 16bit
         checksum with 16bit carry ove rto final 16bit checksum.
RETURNS: Final checksum value
*********************************************************************/
UNSIGNED16 Ccuser_ChecksumFinal(
  UNSIGNED32 last        // last calculated checksum value
)
{
UNSIGNED32 sum1;
UNSIGNED32 sum2;

  // reduce sums (adding hi and low word)
  sum2 = (last >> 16);
  sum1 = (last & 0x0000FFFFul) + sum2;
  sum2 = (sum2 & 0x0000FFFFul) + (sum2 >> 16);
  // return merged sums
  return ((sum2 << 12) | sum1) & 0x0000FFFFul;
}

/*********************************************************************
BOOK:    Section 6.7.1 "Secure message encryption"
DOES:    Encrypts a data block in a secure message
NOTE:    This version NOT optimized for 32 bit architecture
RETURNS: TRUE if encryption completed,
         FALSE if not possible due to parameters
*********************************************************************/
UNSIGNED8 Ccuser_Encrypt(
  UNSIGNED32 *ppad,      // pointer to current one-time pad
  UNSIGNED32 *pdat,      // pointer to the data to encrypt
  UNSIGNED16 first,      // first byte to encrypt
  UNSIGNED16 bytes       // number of bytes to encrypt
)
{
UNSIGNED8 ret_val = FALSE;
#if (Cc_FUNCTIONALITY != Cc_SECFCT_ADVANCED)
UNSIGNED8 lp;
UNSIGNED8 mask;
UNSIGNED8 *p8pad = (UNSIGNED8 *) ppad;
UNSIGNED8 *p8dat = (UNSIGNED8 *) pdat;

  if ( (ppad != NULL) && (pdat != NULL) &&
       (first < Cc_KEY_LEN8) && ((first + bytes) <= Cc_KEY_LEN8)
     )
  { // parameter check ok
    // init mask for looping through key
    mask = (Cc_KEY_LEN8-1);

    for (lp = 0; lp < bytes; lp++)
    { // for length of data
      // XOR data with key
      p8dat[first+lp] ^= p8pad[(first+lp) & mask];
    }
    ret_val = TRUE;
  }
```

```
#else // (Cc_FUNCTIONALITY == Cc_SECFCT_ADVANCED)
UNSIGNED32 buf[4]; // temp buffer needed for encryption

  if ( (ppad != NULL) && (pdat != NULL) && (bytes == 16) )
  {
    memcpy(buf,pdat,16); // copy original data
    AESxxx_Encryption(pdat,buf,ppad);
    ret_val = TRUE;
  }
#endif
  return ret_val;
}


/******************************************************************
BOOK:    Section 6.7.2 "Secure message decryption"
NOTE:    Only used if cryptographic function is not symmetric and
         decryption requires a different function then encryption
DOES:    Decrypts a data block
RETURNS: TRUE if decryption completed,
         FALSE if not possible due to parameters
******************************************************************/
UNSIGNED8 Ccuser_Decrypt(
  UNSIGNED32 *ppad,       // pointer to current one-time pad
  UNSIGNED32 *pdat,       // pointer to the data to decrypt
  UNSIGNED16 first,       // first byte to decrypt
  UNSIGNED16 bytes        // number of bytes to decrypt
)
{
#if (Cc_FUNCTIONALITY != Cc_SECFCT_ADVANCED)
  // use encrypt function, it is symetric
  return Ccuser_Encrypt(ppad,pdat,first,bytes);
#else
UNSIGNED32 buf[4]; // temp buffer needed for encryption
UNSIGNED8 ret_val = FALSE;

  if ( (ppad != NULL) && (pdat != NULL) && (bytes == 16) )
  {
    memcpy(buf,pdat,16); // copy original data
    AESxxx_Decryption(pdat,buf,ppad);
    ret_val = TRUE;
  }

  return ret_val;
#endif
}

/*---------------------- END OF FILE ----------------------------*/
```

Appendix C – Safety-related applications

Applications requiring a safety integrity level (SIL) require certain overhead to be applied to the CAN communication used. There are CAN communication protocols that achieve SIL-3 which for example is used by maritime steer by wire applications. CANopen Safety reaches these safety requirements by duplicating the messages transferred, the second message uses a different CAN message ID and has all data bits inverted. There are specific timing requirements for the message pair, they may not be too long apart to be valid.

Other safety implementations such as Open Safety use a combination of data duplication, time and sequence monitoring and a checksum.

The mechanisms introduced by CANcrypt also increase the safety level of the CAN communication as they also implement several of the basic safety elements required. Looking at the secure messages, CANcrypt uses:

a) Data identification
   Within the message there is a field identifying the data
   (here repeating the CAN ID of the message)
b) Message sequence
   Each preamble also contains a sequence counter
c) Checksum
   A 16-bit checksum field covering a total of 14 bytes
d) Timing requirements
   Preamble and message have a maximum delay, a message
   event/repetition time is not part of CANcrypt but could easily be added
e) Ignoring unsecure messages
   If the requirements of a) to d) are not met, messages are ignored

Missing in CANcrypt is the data duplication. However, as the 16-bit CRC only needs to cover 14 bytes, it is very strong.

The list of published CRC polynomials by Phil Koopman includes a 16-bit CRC that provides a Hamming Distance of eight for up to 15 bytes. Using this polynomial (8fdbh) allows detection of up to seven bit errors in the data transmitted.

At this point we are not claiming that CANcrypt reaches any specific safety integrity level. However, the main mechanisms required are all in place. We currently plan to re-visit this topic in the future.

"There is nothing more visible than what is secret."

"Cousin Cuf"

Appendix D – Message Monitoring

When it comes to monitoring and analyzing CANcrypt messages, then the display of symbols vs. display of plain hex values is a great help. ESAcademy's CANopen Magic already interprets CANcrypt messages.

If you use a different tool but have the options to define message specific symbols, then use the following information when configuring these symbols.

ID	Name	Len	Remarks
171h	CANcrypt configurator message, device 1	3 to 8	1st byte, bit 0-3: request, see 5.3.4 2nd byte, bit 4-7: destination address, device 1 to 15 or 0 for broadcast
172h-17Fh	CANcrypt device X message (X from 2 to 15)	3 to 8	1st byte, bit 0-3: request, see 5.3.4 2nd byte, bit 4-7: destination address, device 1 to 15 or 0 for broadcast
6F0h-6FFh	Bit claiming messages	0	Depending on number of claiming messages only 2 or 16 messages used
6E1h-6EFh	CANcrypt debug message device X (1-15)	8	Custom contents

"Two roads diverged in a wood and I — I took the one less traveled by."

"Bro Rotsfret"

Appendix E - Complete Table of Contents

Olaf Pfeiffer, Andrew Ayre, and Christian Keydel

Embedded Networking with

CAN and CANopen

- Requirements for understanding embedded networking code and communications

- The underlying CAN technology

- Selecting CAN controllers

- Implementation options

- Application-specific examples of popular device profiles

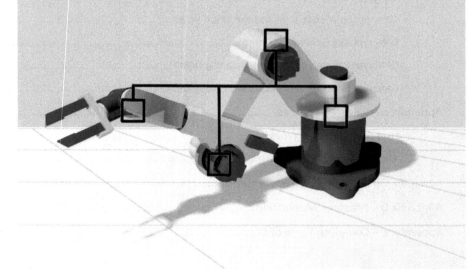

www.canopenbook.com

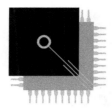

EMBEDDED
SYSTEMS
ACADEMY

Your partner for
Embedded networking technologies

CAN, CANopen, CANcrypt, J1939,
CiA 447, CleANopen, Energybus

Online Shop

Many of our software and hardware products
are also available online - simply point your
browser to

esacademystore.eu

www.ingramcontent.com/pod-product-compliance
Lightning Source LLC
Chambersburg PA
CBHW071114050326
40690CB00008B/1216